Combustion Free Heat, Pow

COMBUSTION-FREE
Heat, Power, and Energy
for Plymouth

David Bricknell with the Power and Energy Group
CLIMATE ACTION PLYMOUTH

November 2022

Climate Action Plymouth – Sustainable Power and Energy Group

Copyright

© 2022

This work is licensed under Creative Commons License
<u>**CC BY-SA**</u>

Acknowledgements

I would like to thank members of the Power and Energy Group of Climate Action Plymouth: Didier Cavrot, Gay Jones, Ricky Lowes, Simon Pannett, Zoë Smith, and Malcolm Teague for their comments on drafts of this book and for their support and encouragement throughout the process of writing.

With thanks to Friends of the Earth for funding the printing of this book.

"..acting on climate is not being restricted by a lack of scientific knowledge or technological options, but by entrenched power structures and an absence of political will." IPCC [1]

There is no credible pathway to 1.5°C in place. The failure to reduce carbon emissions means the only way to limit the worst impacts of the climate crisis is a rapid transformation of societies. – UN Environment Agency – Emissions Gap Report 2022: The closing window – climate crisis calls for rapid transformation of societies.

"Technology is going to be an important part of dealing with this, but it is not a magic solution and isn't going to deal with it on its own. If you take 2050, any technology that you can't see already working is not going to save our bacon because of the scale of which these things need to be introduced." Patrick Vallance [2]

The World Meteorological Organization reports [3]:

- Carbon dioxide, methane and nitrous oxide are at record levels in the atmosphere as emissions continue. The annual increase in methane, a potent greenhouse gas, was the highest on record.
- The sea level is now rising twice as fast as 30 years ago, and the oceans are hotter than ever.
- Records for glacier melting in the Alps were shattered in 2022, with an average of 4 metres in height lost.
- Rain (not snow) was recorded on the 3,200m-high summit of the Greenland ice sheet for the first time.
- The Antarctic sea-ice area fell to its lowest level on record, almost 1m km² below the long-term average.

Abbreviations

AGR	Advanced Gas cooled Reactor	Toe	kilo tonnes oil equivalent
AMR	Advanced Modular Reactor	kW	kilo Watt
ASHP	Air Source Heat Pump	kWh	kilo Watt hour
BESS	British Energy Security Strategy	LED	Light Emitting Diode
BESS	Battery Energy Storage System	LCCC	Low Carbon Contracts Company
BEV	Battery Electric Vehicle	LCOE	Levelised Cost of Electricity
BTU	British Thermal Unit	LNG	Liquified Natural Gas
CAP	Climate Action Plymouth	LPG	Liquid Petroleum Gas
CCGT	Combined Cycle Gas Turbine	LTO	Lithium Titanate Oxide
CCUS	Carbon Capture, Utilisation, and Storage	Ma	Millions of years ago
CEAP	Climate Emergency Action Plan	MSR	Molten Salt Reactor
CfD	Contracts for Difference	mtCO2e	Million tonnes of Carbon Dioxide equivalent
CH	Hydrocarbons	MW	Mega Watt
CH4	Methane	MWh	Mega Watt hour
CHP	Combined Heat and Power	NDC	Nationally Determined Contributions
CO	Carbon Monoxide	NMHC	Non-Methane Hydro-Carbon
COP	Conference of Parties	N_2O	Nitrous Oxide
CoP	Coefficient of Performance	NOx	Nitrogen Dioxide (principally)
CO_2	Carbon Dioxide (GWP=1)	OFGEM	Office of Gas and Electricity Markets
CO_{2e}	Carbon Dioxide equivalent	ODP	Ozone Depletion Potential
COMEAP	Committee on the Medical Effects of Air Pollution	PEC	Plymouth Energy Community
		PFC	Perfluorocarbons
COPD	Chronic Obstructive Pulmonary Disease	PCC	Plymouth City Council
CPRE	The countryside charity	PCC CEAP	Plymouth City Council Climate Emergency Action Plan
Cu.ft	cubic feet		
Cu.m	cubic metres	PWR	Pressurised Water Reactor
dBA	decibel measured with 'A' Scale	PM	Particulate Matter
DNO	Distribution Network Operator	PV	Photo Voltaic
DUKES	Digest of UK Energy Statistics	REG	Regional Energy Grid
ECO	Energy Company Obligation	RSPB	Royal Society for the Protection of Birds
EPC	Energy Performance Certificate	SCoP	Seasonal Coefficient of Performance
EPR	European Pressurised Water Reactor	SEDBUK	Seasonal Efficiency of Domestic Boilers in the UK
EPS	Expanded Polystyrene		
ErP	Energy related Products	SF6	Sulphur Hexafluoride
EU	European Union	SOx	Sulphur Oxides
EV	Electric Vehicle	SMR	Small Modular Reactor
FES	Future Energy Scenario	SPA	Special Protection Area
F Gases	Fluorinated gases	SSR	Stable Salt Reactor
GB	Great Britain	toe	tonnes of oil equivalent
gCO_2/kWh	grammes of Carbon Dioxide per kWh	tCO_2e	tonnes of Carbon Dioxide equivalent of
GEL	Geothermal Engineering Ltd	TWh	Terra Watt hours
GHGs	Green House Gases	UK	United Kingdom Great Britain and Northern Ireland
GHNF	Green Heat Network Fund		
GIR	Green Industrial Revolution	VAT	Value Added Tax
GSHP	Ground Source Heat Pump	V2G	Vehicle to Grid
GW	Giga Watt	µm	micro metre
GWh	Giga Watt hours	UDDGPP	United Downs Deep Geothermal Power Project
GWP	Global Warming Potential		
HNIF	Heat Networks Investment Fund	UK	United Kingdom
HFC	Hydrofluorocarbons	USA	United States of America
HP	Heat Pump	WHO	World Health Organisation
kWh	Giga Watt hours	WSHP	Water Source Heat Pump
kWp	kilo Watts Peak	ZEB	Zero Emission Boiler
H&BS	Heat and Buildings Strategy		
HFC	Hydrofluorocarbons		
HP	Heat Pump		
IPCC	Intergovernmental Panel on Climate Change		

Definitions, Units and Measurements

Power - **Watt**, kW, horsepower, BTU/h

Energy - **Joule**, kWh, BTU, calorie, toe.

One **tonne of oil equivalent** (toe) = 11,630kWh

Green House Gases (CO_2 **equivalent) and Global Warming Potential**

Tonnes of Carbon Dioxide equivalent - tCO_2e

Toxic Emissions - Principally those resulting from combustion

Nitrogen Dioxide NO_2, Particulates (PM_{10}, $PM_{2.5}$, PM_1, $PM_{0.1}$)

Scale:
kilo -	1,000
Mega -	1,000,000
Giga -	1,000,000,000
Terra -	1,000,000,000,000
Peta -	1,000,000,000,000,000

Unit Conversion:
1 BTU = 0.000293 kWh
1,000 cu.ft natural gas = 1 m BTU = 1GJ = 278kWh
1,000 cu.m natural gas = 36.9 m BTU = 10,814kWh

Carbon Definitions

Carbon Neutral is achieved by reducing CO_2e to zero through offsetting, including through buying carbon credits.

Net-Zero Carbon is achieved by matching the amount of CO_2e emitted with CO_2e removed from the atmosphere by carbon removal or by carbon capture and storage. This cannot be achieved by offsetting.

Zero Carbon is when no CO_2e is emitted.

Negative Carbon is when the process reduces carbon in the environment.

Relevant Documents

IPCC

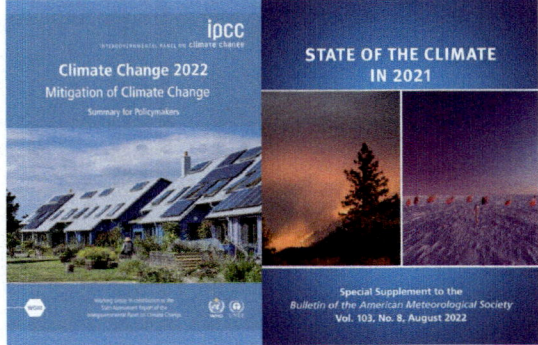

UK Government

Plymouth City Council

Combustion-Free Heat, Power, and Energy

Table of Contents

COPYRIGHT I
ACKNOWLEDGEMENTS II
ABBREVIATIONS IV
DEFINITIONS, UNITS AND MEASUREMENTS V
CARBON DEFINITIONS VI
RELEVANT DOCUMENTS VII

SUMMARY ....- 1 -

ACTIONS ....- 3 -

1. INTRODUCTION ....- 4 -

2. THE CLIMATE EMERGENCY AND GREENHOUSE GAS EMISSIONS ....- 6 -

CLIMATE EMERGENCY ....- 6 -
MASS EXTINCTIONS ....- 7 -
GREENHOUSE GASES (GHG) ....- 10 -
GREENHOUSE GAS SOURCES ....- 13 -
ENERGY – END USE (2019) [BEIS ENERGY CONSUMPTION IN THE UK] ....- 15 -
LIFECYCLE GREENHOUSE GAS EMISSIONS ....- 16 -

3. TOXIC EMISSIONS ....- 17 -

TOXIC EMISSIONS ....- 17 -
EMISSION STANDARDS ....- 18 -

4. THE ENERGY MARKET, FUEL TAXATION, LEVIES, CONTRACTS FOR DIFFERENCE ....- 19 -

ENERGY MARKET ....- 19 -
UTILITY BILLS ....- 20 -
ENVIRONMENTAL AND SOCIAL LEVY ....- 20 -
CONTRACTS FOR DIFFERENCE ....- 22 -

5. ENERGY USE/DEMAND REDUCTION ....- 23 -

UK GOVERNMENT STRATEGY AND INITIATIVES ....- 23 -
PCC'S CEAP INITIATIVES ....- 24 -
HOME ENERGY USE ....- 24 -
SPACE HEATING ....- 25 -
DOMESTIC ELECTRICAL USAGE ....- 26 -
ENGLAND'S HOUSING STOCK ....- 27 -
INSULATION MEASURES, INCLUDING DOUBLE GLAZING. ....- 28 -
VENTILATION ....- 31 -
PASSIVHAUS ....- 32 -
INDUSTRIALISING BUILDING INSULATION – ENERGIESPRONG ....- 32 -

Climate Action Plymouth – Sustainable Power and Energy Group

RETROFITTING A VICTORIAN TERRACED HOUSE .... - 32 -
PLYMOUTH'S HOUSING STOCK .... - 34 -
PLYMOUTH'S PROGRESS WITH CARBON SAVING ON MUNICIPAL BUILDINGS .... - 34 -

6. HEATING AND HOT WATER .... - 36 -

UK GOVERNMENT STRATEGY AND INITIATIVES .... - 36 -
PCC's CEAP INITIATIVES .... - 37 -
HEATING AND HOT WATER .... - 37 -
DOMESTIC BOILERS .... - 38 -
GAS/OIL/SOLID FUEL BOILER .... - 38 -
ELECTRIC BOILERS .... - 40 -
HEAT PUMPS .... - 40 -
THE ZERO EMISSION BOILER .... - 45 -

7. ELECTRICITY GENERATION, DISTRIBUTION, AND STORAGE .... - 46 -

UK GOVERNMENT STRATEGY AND INITIATIVES .... - 47 -
PCC's CEAP INITIATIVES .... - 47 -
ELECTRICITY POWER GENERATION .... - 47 -
REGIONAL ENERGY GRIDS .... - 53 -
LEVELISED COST OF ELECTRICITY (LCOE) .... - 54 -
ELECTRICITY INTERCONNECTORS .... - 55 -
ENERGY STORAGE .... - 57 -

8. FOSSIL FUELS AND BIOMASS .... - 62 -

UK GOVERNMENT STRATEGY AND INITIATIVES .... - 63 -
NATURAL GAS .... - 63 -
WOOD BURNING AND BIOMASS .... - 68 -
CARBON CAPTURE UTILISATION AND STORAGE (CCUS) .... - 70 -

9. RENEWABLES (ZERO CARBON IN OPERATION) .... - 71 -

UK GOVERNMENT STRATEGY AND INITIATIVES .... - 72 -
PCC's CEAP INITIATIVES .... - 72 -
NON-COMBUSTION, ZERO CARBON .... - 72 -
WIND .... - 75 -
SOLAR PHOTO VOLTAIC (PV) .... - 81 -
HYDRO .... - 84 -
TIDAL STREAM .... - 87 -
WAVE POWER .... - 89 -
GEOTHERMAL .... - 91 -

10. NUCLEAR-FUELLED STEAM POWER GENERATION .... - 93 -

UK GOVERNMENT STRATEGY AND INITIATIVES .... - 93 -
NUCLEAR POWER PLANT IN UK .... - 94 -

LARGE-REACTOR CAPABILITY	- 95 -
SMALL MODULAR REACTORS (SMR)	- 96 -
ADVANCED MODULAR REACTORS (AMR)	- 97 -
NUCLEAR ENERGY (FISSION) SECURITY, ENVIRONMENTAL RISKS, AND CONCERNS	- 98 -
FUSION	- 100 -
INNOVATION IN NUCLEAR	- 101 -

11. DISTRICT HEATING ....- 102 -

UK GOVERNMENT STRATEGY AND INITIATIVES	- 102 -
PCC'S CEAP INITIATIVES	- 102 -
DISTRICT HEATING WORLDWIDE	- 102 -
ENVIRONMENTAL IMPACT OF DISTRICT HEATING	- 104 -

12. HYDROGEN ....- 105 -

UK GOVERNMENT STRATEGY AND INITIATIVES	- 105 -
LOCAL DEVELOPMENTS	- 105 -
HYDROGEN FOR HOME HEATING	- 105 -
HYDROGEN IN A FUEL CELL	- 106 -
HYDROGEN PRODUCTION	- 106 -
HYDROGEN FOR HEATING	- 107 -
HYDROGEN HEATING EFFICIENCY.	- 107 -

13. CONCLUSIONS ....- 108 -

14. ABOUT THE AUTHOR AND CLIMATE ACTION PLYMOUTH ....- 111 -

15. REFERENCES ....- 112 -

x

SUMMARY

Climate change is real and is happening now. Unless radical action is taken to curb greenhouse gas emissions to levels advised by climate scientists, rising temperatures across the globe will lead to more weather extremes of heat, cold, rain and drought, the consequences of which will be mass starvation, mass migration, biodiversity decline, and another mass extinction.

Climate change is caused by increasing greenhouse gas emissions which act to prevent heat from escaping the atmosphere. Greenhouse gas emissions principally arise from the combustion of fossil fuels. Fossil fuels power most of our transport and most of our heating and hot water. Electrifying transport is a revolution already underway; electrifying heating and hot water has yet to make progress. Electricity generation has decarbonised but there is still a long way to go.

The UK's commitment to net-zero by 2050 requires strong and decisive actions now. The absolute reduction in our carbon emissions by 2025, rather than simply reducing the rate at which emissions are increasing, needs to start urgently. Without this action, net-zero targets of 2050 will not be met. For the sake of the planet, and the human race, the 1.5°C target must be met or bettered.

Environmental and social levies are disproportionately applied to electricity bills with little being applied to gas. This must change if the electrification of heat is to be successful and gas boilers are to be discouraged.

The most optimistic forecast of future energy demand by the National Grid requires a reduction in annual energy usage of some 30%. Improving the energy efficiency through better levels of insulation is essential if there is to be any hope of achieving this reduction.

Even with improved energy efficiency, to electrify heating will require a fourfold increase in electricity generation. If electricity is to be generated solely from combustion-free, zero carbon sources, then an order of magnitude increase in zero-carbon non-combustion renewables is required.

Gas is not a solution. Recent energy security issues have highlighted dependence on natural gas for home heating and hot water as well as for electricity generation. Pipeline gas from the North Sea is an international resource, sold on the open international market and not preferentially to the domestic market. Liquified Natural Gas (LNG) is considerably worse for carbon emissions than pipeline gas.

Continued use of combustion fuels cannot be part of the solution. Future combustion of fuels relies upon Carbon Capture, Utilisation and Storage (CCUS), a technology which is a long way from being economically viable.

The current high LCOE (Levelised Cost of Electricity) for fossil fuel derived electricity may be helpful in moving to zero emission renewable sources of energy. Levelised cost of electricity is a measure of the average net present cost of electricity generation for a generator over its lifetime. It is used to compare different methods of electricity generation on a consistent basis.

Of the zero emission renewables, wind and solar are proven technologies with an attractive LCOE. Other zero emission renewables, hydro, tidal, wave, and geothermal, are either at a low technology readiness, not scalable, or have too high an LCOE, although all should continue to be funded at a low level and their technology readiness monitored.

Energy storage significantly enhances the load factor for all intermittent renewables. As with insulation, it is a technology which is available and is essential to growing renewable power generation.

Nuclear energy is zero carbon in generation and low carbon over its lifetime. New nuclear is expensive, slow to bring on-line, and does not complement zero carbon intermittent renewables. Small modular reactors may address the speed to generation and maybe the LCOE, but all types of nuclear power have the waste disposal and fuel sourcing issues. Stable Salt Reactors (SSR) using existing nuclear waste may be a viable and complementary technology to intermittent renewables but is unlikely to be available in the short timescales required.

Hydrogen can contribute to decarbonisation but to do so it must be produced from zero carbon renewable electricity and should not be used in combustion. Hydrogen will be a higher cost than renewable electricity and will likely be a higher cost than current fossil fuels.

To decarbonise heating, power, and energy rapidly:

- **Stop combustion (burning) of fossil fuels and biomass.**
- **Redistribute taxes and levies from clean electricity onto gas.**
- **Build new dwellings, and update existing dwellings, to a high standard of insulation and ventilation to reduce heating requirements and to ensure heat pumps are an affordable solution.**
- **Replace natural gas boilers by electric heating / heat pumps.**
- **Generate electricity by technology-ready and low-cost, zero-emission, wind, solar, and hydro renewables, backed by grid-level energy storage.**
- **Incentivise domestic solar energy generation and storage.**
- **Seed fund future energy solutions including wave, tidal, and geothermal.**
- **Invest in local energy schemes, where feasible.**

Actions

PROPOSED GOVERNMENT ACTIONS

- Eliminate subsidies for fossil fuel exploration and production and transfer them to renewables.
- Refuse further fossil fuel exploration licenses.
- Move the environmental and social levy from electricity to gas.
- Reduce energy usage by a programme of insulation and home efficiency improvements.
- Increase electricity production share from combustion free sources, particularly wind, and solar.
- Eliminate biomass.
- Seed fund innovation in hydro, tidal, wave, geothermal and stable salt reactors.
- Invest in the National Grid and regional power distribution to ensure they are prepared for the energy transition.
- Remove incentives for new and replacement gas boilers and increase incentives for a transition to heat pumps.

PROPOSED LOCAL AUTHORITY ACTIONS

- Mandate all new build dwellings have net zero energy usage.
- Ensure all new dwellings have solar panels backed by energy storage.
- Improve insulation and ventilation for all domestic dwellings with the aim of an energy band of A or the nearest feasible standard, enabling the elimination of combustion based heating and hot water, or, as a last resort, demolish and re-build.
- Improve the carbon efficiency of all local authority buildings.
- Identify and make available, locations for suitable solar and wind farms.

PERSONAL CONTRIBUTION

- Improve the insulation of your property to reduce energy demand.
- Electrify your heating and cooking - change your gas boiler and gas hob for either heat-pump or direct electric.
- Upgrade 'wet' and 'cold' appliances to more efficient, lower energy models.
- Move electricity usage to off-peak low-carbon supply, avoiding peak 4-7pm slot.
- Reduce boiler flow temperature to 50°C (combi) or 60°C (system with tank).
- Lower the thermostat on your heating, wear warmer clothes, take shorter showers.
- Use a 100% renewable electricity supplier.

1. Introduction

1.1 This document sets out the views of Climate Action Plymouth (CAP) on heat, power, and energy in the UK and, more specifically, in Plymouth.

1.2 There is a growing realisation amongst the public as well as politicians that climate change is real, the effects of which are happening now. Climate change is causing irreversible temperature increases across the globe leading to more weather extremes of heat, cold, rain, and drought, the consequences of which will be mass starvation, mass migration, biodiversity decline, and ultimately another mass extinction.

1.3 Climate change is a result of the increase in greenhouse gases, particularly, carbon dioxide. Carbon dioxide emissions are largely a product of the combustion of fossil fuels which are used to heat homes and to produce electricity.

1.4 Consideration is given to the recent Government strategies: British Energy Security Strategy [4]; and Heat and Buildings Strategy October 2021 [5]. The document acknowledges and comments on the above documents. The recent political turmoil seems to have settled down with a return to the policies of earlier in 2022.

1.5 Plymouth City Council acknowledges the climate emergency and has produced three Climate Emergency Action Plans, the latest of which is the 2022 plan [6]. These plans are considered when setting out CAP's objectives.

1.6 There are some important definitions and terminologies that are often confused when discussing 'emissions', and 'carbon'. These are defined first.

1.7 The causes of the climate emergency, greenhouse gases, and toxic emissions from burning fossil fuels, are discussed in Chapters 2 and 3. These set the scene for the necessary actions and identify where a misunderstanding of the mechanisms of climate change may lead to an opposite outcome than that intended.

1.8 Within the UK, the most significant contributors of greenhouse gas emissions are those produced by transport in burning oil products (petrol and diesel) and those produced by heating and hot water from burning natural gas and solid fuels.

1.9 This book therefore tackles the issues around heating and hot water and considers the zero-carbon, sustainable and renewable options, their affordability, their technology readiness, and their potential impact on greenhouse gas within the short timescales available to affect the direction of climate change. Transport is covered by CAP's "Combustion-free mobility for Plymouth". Agriculture and Industry will be covered by other CAP publications.

1.10 Economics and affordability play a significant part in what energy we consume whilst "Contracts for Difference" (CfD) affects the type of electricity that is generated. These are discussed in chapter 4.

1.11 Taxation and levies are a significant tool for encouraging the outcomes we want and penalising those we do not want. Currently our taxation framework is a decade out of date, encouraging gas usage and penalising electricity use. Consequently, the fuel poor are disproportionately impacted by the misdirected levies.

1.12 Chapter 5 looks at energy efficiency and insulation and the potential to reduce demand. Improved energy efficiency will ensure that renewable energy is able to meet the country's demand without the need for fossil fuels.

1.13 Chapter 6 examines heating/hot water energy usage, gas boilers efficiency and the alternative of heat pumps.

1.14 Chapter 7 considers Great Britain's electricity generating capability, what it needs to do to deliver a net-zero 2050, the benefits of interconnectors to our electricity grid, and the importance of energy storage.

1.15 Chapters 8 takes a detailed look at the fossil fuels, natural gas, and solid fuels (wood and biomass) as well as Carbon Capture, Utilisation and Storage.

1.16 A zero-carbon future requires widespread electrification for both transport and heating/hot water. The required transformation of electric generation from one where over half is produced by the combustion of fossil fuels to one of zero-emission sustainable and renewable technologies is covered in Chapter 9. These technologies include wind, solar, hydro, tidal, wave, and geothermal.

1.17 Nuclear energy is showing signs of a return with plans for more large reactors and many more 'small modular' reactors. Nuclear has strengths but is expensive and slow to bring to market, as well as having security and waste disposal issues. Nuclear is considered in Chapter 10.

1.18 Chapter 11 considers district heating, something that is currently largely delivered with the burning of fossil fuels rather than by using renewable energy.

1.19 Hydrogen has captured the attention of many looking for a simple swap in fuel for natural gas. Hydrogen will have a role in energy storage, but it comes with high cost and many drawbacks. These are considered in Chapter 12.

1.20 This book aims to:
- increase understanding around energy use towards a zero-climate impact heat and energy plan for Plymouth.
- produce publications which are accessible to policymakers and laypeople as a reference guide and a campaign pamphlet
- influence heat and energy policy in Plymouth
- engage Plymouth citizens in the discussion and gather support for lobbying
- support the wider climate movement by sharing book and pamphlet widely with other climate groups for broader distribution of ideas.

2. The climate emergency and greenhouse gas emissions

Greenhouse gases keep the earth's atmosphere warm. Increasing Greenhouse gas emissions cause the earth's atmosphere to heat up.

Greenhouse gas emissions are primarily a result of the burning of petroleum for transport and burning natural gas for heating.

Average annual global greenhouse gas emissions are at their highest levels in human history. [7]

The human-race faces an existential crisis if it does not contain the global temperature rise to below 1.5°C. Without immediate and deep emissions reductions, containing global temperature rise to 1.5°C is beyond reach. Currently, with existing Nationally Determined Contributions (NDC), the earth is headed for 2.5°C or greater.

Lifecycle emissions of all renewables and nuclear are much lower. Fossil-fuel combustion is the largest emitter of greenhouse gases.

Widespread electrification from renewables together with improved insulation and energy efficiency are vital to reduce greenhouse gases.

The UK Government's *British Energy Security Strategy* (BESS) [8] **aims for a 40% reduction in natural gas use by 2030 but has a new oil and gas licensing round for 2022. Domestic gas production will remain a core part of UK energy demand into the 2030s. This is not consistent with an immediate and deep emissions cut.**

Climate emergency

2.1 The climate emergency is the crisis faced by all humanity caused by the accelerating increase in average global temperatures. Global warming is a result of the increase in greenhouse gases from transport, energy, business, residential, industrial, and agriculture. (Shipping and aviation are not captured in official statistics). [9]

2.2 The speed of global temperature increase caused by the rapidly increasing levels of greenhouse gases has not been previously experienced throughout human history. "The scientific evidence is unequivocal: climate change is a threat to human wellbeing and the health of the planet. Any further delay in concerted global action will miss a brief and rapidly closing window to secure a liveable future." Hans-Otto Pörtner, co-chair of working group of the IPCC.

2.3 "Extreme heatwaves, floods and droughts, far outside the normal tolerances of cities and towns, will destroy lives and livelihoods globally. The knock-on effects of crop failures, migration and economic disruption will overload our political institutions and damage our ability to respond to unfolding events." [10]

2.4 Greenhouse gas concentrations, global sea levels and ocean heat content reached record highs in 2021 (32nd annual State of Climate report [11]).

2.5 "(Plymouth City) Council has declared a climate emergency". This means that we [PCC] are working with our partners to make Plymouth **'carbon neutral'** by 2030. The 2030 target stems from the Intergovernmental Panel on Climate Change (IPCC) 2018 report which outlined the impact of a temperature rise of more than 1.5°C". The report stated that urgent action was required to tackle climate change. [12] CAP applauds PCC's choice of 2030 rather than 2050 for carbon neutrality.

Mass extinctions

2.6 Evidence exists for global temperatures and atmospheric CO_2 over the past 500 million years (Ma). During that time global temperatures have varied, and sea levels have changed.

2.7 There have been five major extinctions during that time: Ordovician-Silurian (443Ma), Late-Devonian (364MA), Permian-Triassic (251Ma), Triassic-Jurassic (200Ma) – Cretaceous-Paleogene (65Ma). All are associated with global warming. [13 14 15]

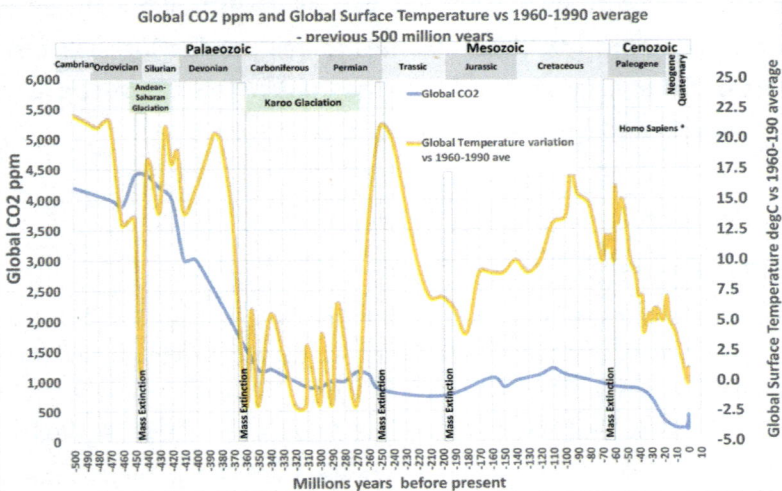

Global temperature and CO_2 levels over the previous 500 million years.

2.8 **Holocene/Anthropocene Extinction** Today we are entering another mass extinction – this time caused by human activity, principally the burning of fossil fuels causing the rapid release of previously buried carbon as greenhouse gases into the atmosphere.

2.9 For the past 300,000 years the earth's CO_2 has been relatively stable. During this period Homo Sapiens evolved. The Holocene (~10,000years BC) saw humans settle, moving from hunter-gatherers to domestication of plants and animals. Humans have had about 400 generations since then in which to evolve.

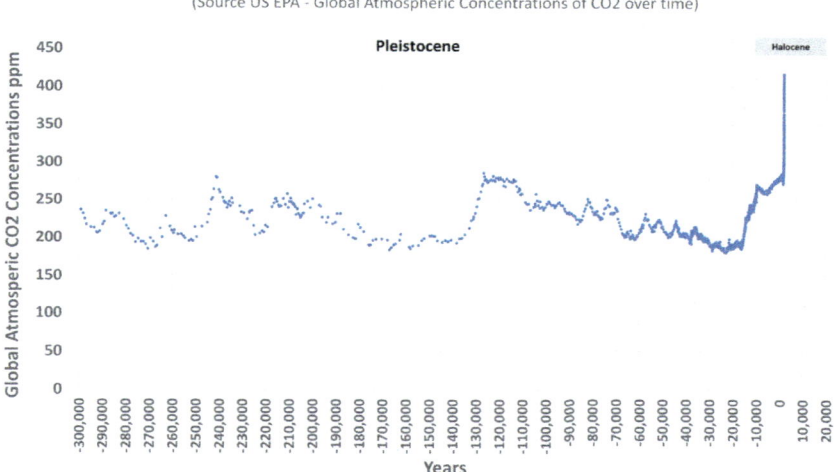

CO_2 concentrations over the past 300,000 years showing clearly to sharp increase following the industrial revolution.

2.10 During the late eighteenth to early nineteenth century, the industrial revolution began, fuelled by the burning of hydrocarbons to produce energy – initially coal, then oil, and now methane (natural gas).

2.11 Burning (combustion) of hydrocarbons (fossil fuels) produces carbon dioxide CO_2 as a by-product. Scientists from as early as the beginning of the nineteenth century through to today have studied and documented the warming effect on the Earth of trapping infra-red solar radiation beneath a blanket of GHGs. [16]

2.12 The increase in atmospheric greenhouse gases since the beginnings of the industrial revolution is measurable, and its effects, in terms of rising air and sea water temperatures, are clear and measurable. The internationally recognised CO_2 measurements are taken by the Scripps Institution of Oceanography, UC San Diego at the Mauna Loa Observatory, Hawaii. [17]

Climate Action Plymouth – Sustainable Power and Energy Group

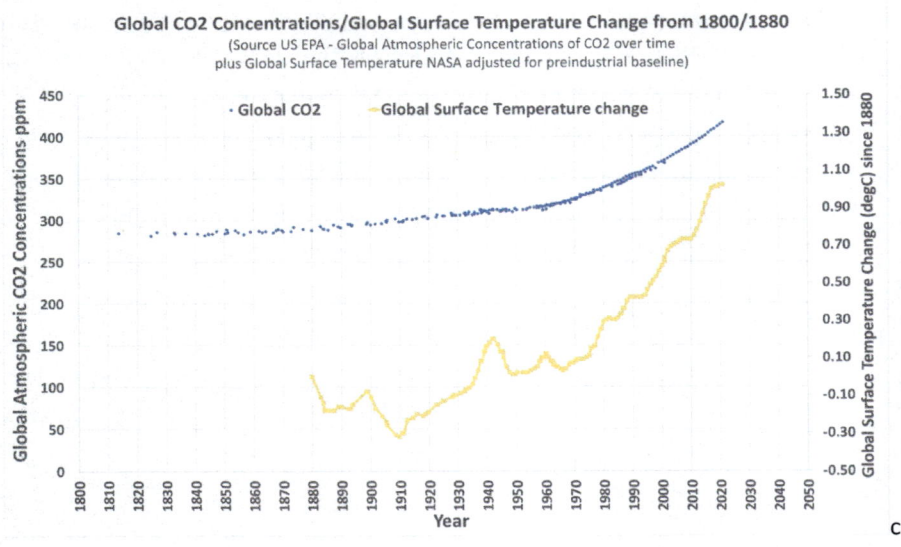

CO₂ increase and Global Surface Temperature change from 1880.

2.13 The sixth assessment of the Intergovernmental Panel on climate change was summarised by Secretary-General Antonio Guterres as nothing less than 'a code red for humanity. The alarm bells are deafening, and the evidence is irrefutable'. [18][19]

2.14 To return to a stable climate enabling the continued existence of the human race and other species within a bio-diverse environment, we must reduce atmospheric greenhouse gases, that is, we must **stop burning fossil fuel**.

2.15 We must ensure that greenhouse gases in 2025 (3 years) are at their peak and that within eight years (2030) greenhouse gases are sufficiently reduced to prevent a rise of over 1.5°C. [20]

2.16 A rise of 1.5°C will see an ice-free arctic, half of land and sea creatures severely affected, and a greater frequency of deadly heatwaves and severe flooding. Many of these effects are evident in 2022 and things will continue to get worse if action is not taken urgently. [21]

2.17 On our current trajectory, the rise to between 2.6 to 3.9°C would be catastrophic for the world as we know it – a world unliveable in by humans as we know them today. [22]

Greenhouse gases (GHG)

2.18 The 'Greenhouse' effect is well documented and is accepted scientifically. **Greenhouse gases** increase the amount of radiation trapped and hence the amount that the planet warms. This is known as the greenhouse effect and is the cause of global warming.

2.19 The Earth receives radiation from the sun (sunlight), some of which is absorbed and some of which is lost into space. When the radiation is reflected back to the Earth rather than escaping to space, it raises the temperature of the planet.

2.20 The composition of the atmosphere determines how much of the radiation received from the sun is lost back into space and how much is trapped in the Earth's atmosphere by greenhouse gases. The more greenhouse gases in the atmosphere, the more of the Sun's heat is trapped in the atmosphere.

2.21 Greenhouse gases include seven main gases – carbon dioxide, methane, nitrous oxide, and four F-gases. If hydrogen is included, then there would be eight.

Combustion of fuel related

- Carbon Dioxide (CO_2) GWP 1
- Methane (CH_4) GWP x25 (25 times the same mass of CO_2 over 100 years)
- Nitrous Oxide (N_2O) GWP x298

Fluorinated gases (F-gases)

- Hydrofluorocarbons HFC GWP x 12-14,800
- Perfluorocarbons PFC GWP x 7,300 – 17,340
- Sulphur Hexafluoride SF6 GWP x 22,800
- Nitrogen Trifluoride GWP x17,200

Other gases

- Hydrogen GWP x 11 [23]

2.22 **Global Warming Potential** (GWP) measures the impact of the same mass of each gas relative to carbon dioxide CO_2 over a 100-year timescale. This enables a carbon dioxide equivalent CO_2e to be assessed. Note: Potential is used in the scientific sense: how much greater is the warming it *will* cause – not that it *might* cause but that it will cause!

2.23 Carbon dioxide, methane, and nitrous oxide are all emitted during the combustion of fuels. [24]

2.24 Carbon dioxide is by far the greater from combustion but both methane and nitrous oxide depend not just on the fuel type but also on the combustion technology.

2.25 Methane is emitted during the production, transport, and combustion of fossil fuels, rotting waste in landfill, livestock, and wetlands.

2.26 Nitrous oxide, also known as laughing gas, is both a greenhouse gas and an ozone depleting gas. It is mostly emitted into the atmosphere from agriculture (primarily nitrogen fertiliser), combustion of fossil fuels as well as industrial processes. [25] Nitrous oxide has also been found to be a by-product of lean burn combustion of natural gas enriched by hydrogen. [26]

2.27 Fluorinated gases are emitted from a variety of manufacturing and industrial processes. As well as being GHGs they are also ozone depleting.

2.28 Hydrogen is given a GWP because leakage of hydrogen into the atmosphere will reduce the concentration of hydroxyl radicals, the major tropospheric oxidant. These hydroxyl radicals will then not be available to deal with methane, meaning methane will have a longer atmospheric lifetime. Overall then the GWP is greater than if hydrogen wasn't produced. [27]

2.29 Limiting global warming to around 1.5°C requires GHG emissions to peak before 2025 and to be reduced by 43% by 2030.

2.30 Greenhouse gas emissions worldwide have reached a new high in 2021 and continue to increase into 2022.

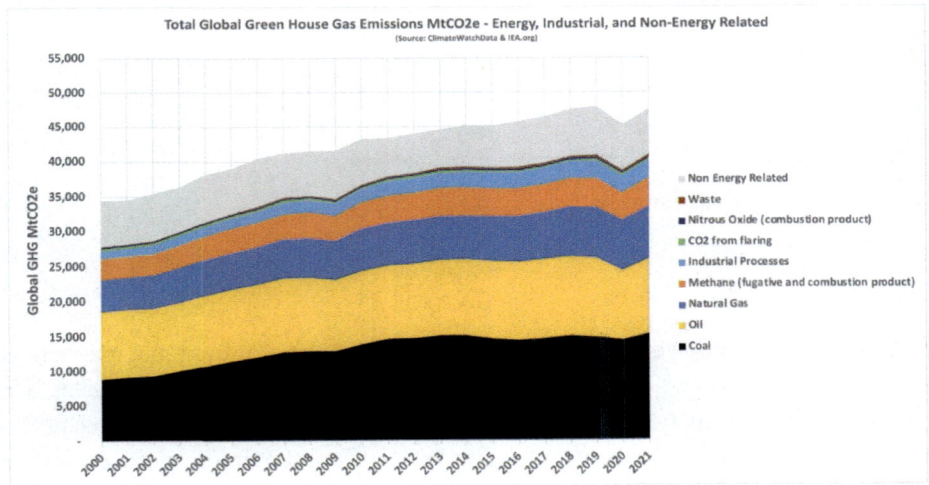

Global greenhouse gas emissions continue to increase into 2021

2.31 The UK has made significant steps in reducing territorial greenhouse gas emissions – a 39% reduction since 1990. [28]

Combustion-Free Heat, Power, and Energy

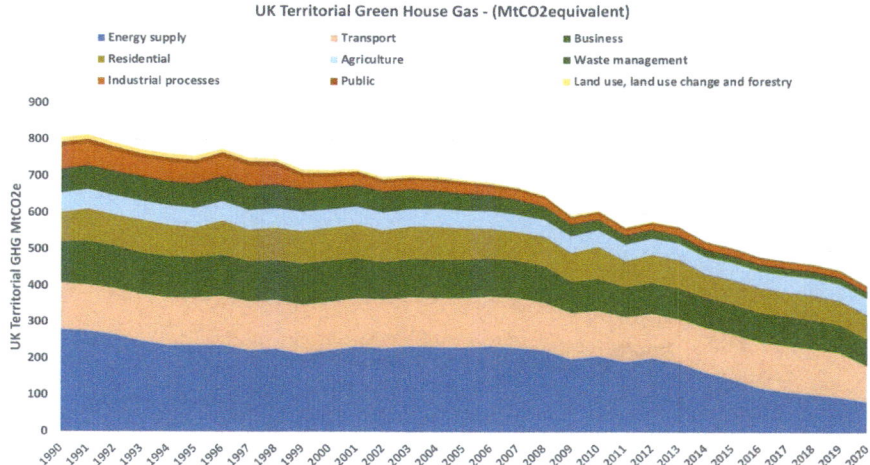

The UK has made significant reductions in GHG emissions since 1990, mostly with energy generation moving away from coal from the mid-eighties.

2.32 This has come about largely from a change from manufacturing to service-based industries.

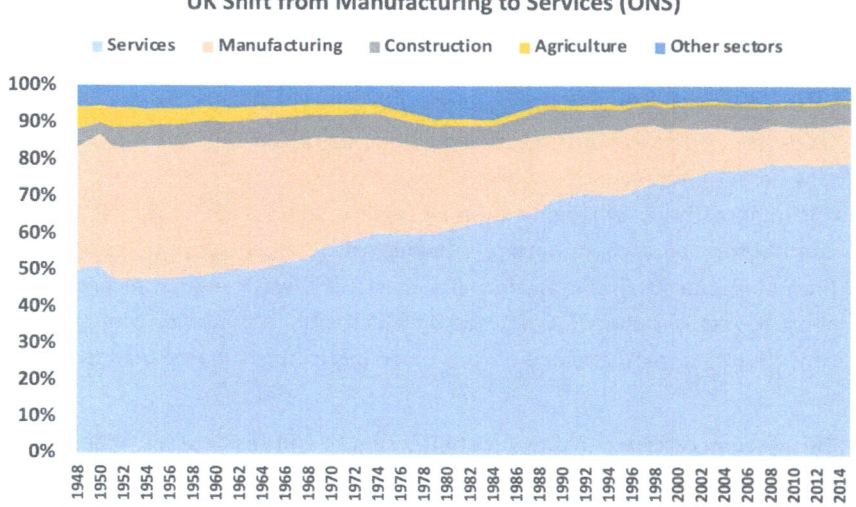

The UK's move from manufacturing into services has contributed to the reduction in greenhouse gas emissions –shows composition of labour force changing with time.

2.33 A country's emissions should also include those of the manufacturing country. Much of the reduction in manufacturing has been offset by imported CO_2 emissions from our increased reliance on goods from China. [29]

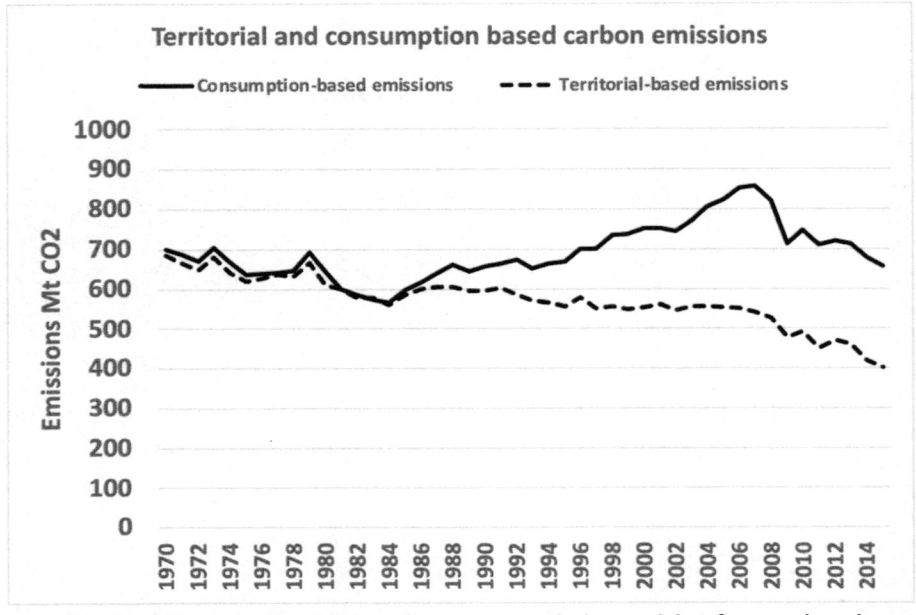

Much of the reduction in greenhouse gas emissions arising from reduced manufacturing has been 'exported' to other countries.

Greenhouse gas sources

2.34 UK territorial greenhouse gases emissions comprise the four main types – carbon dioxide CO_2, methane CH_4, nitrous oxide N_2O, and fluorinated gases F Gases. The UK documents these emissions in **nine major sources** of which six are the major contributors of carbon dioxide - **transport, energy supply, residential, business** (including industrial combustion and electricity, etc.), **industrial processes** (such as cement, etc.), **public** (public sector electricity, etc.), and three are the major contributors of methane and nitrous oxide (**agriculture, waste management,** and **land use**).

2.35 This document concentrates on energy supply and residential. Transport is covered by CAP's publication "Combustion Free Mobility for Plymouth". Business and industrial, and agriculture, waste and land use change will be explored in further publications.

Combustion-Free Heat, Power, and Energy

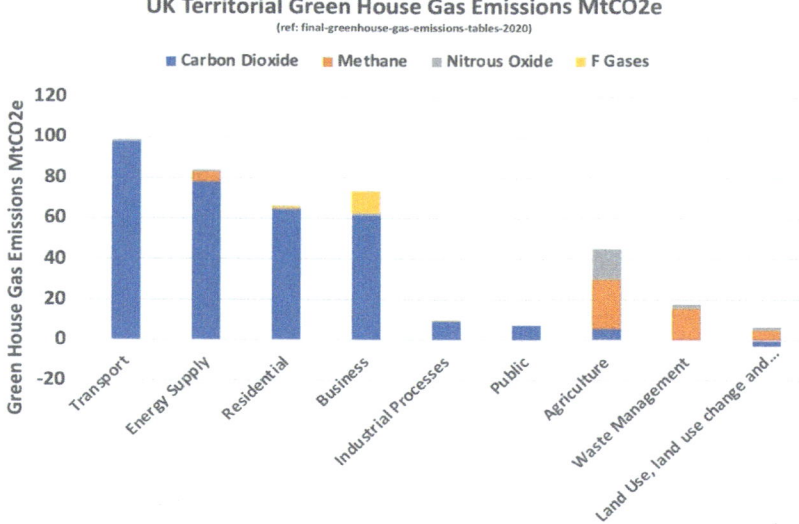

2020 UK territorial GHG emissions by source and by greenhouse gas emission

2.36   The UK Government provisional emissions data provides a source for the UK's 2020 annual greenhouse gases emission of 405.5 million tonnes of carbon dioxide equivalent. [30]

2.37   Carbon dioxide arises from the combustion of gaseous fuels (predominantly natural gas), petroleum (petrol, diesel, and other petroleum gases), coal, and other solid fuels, such as wood.

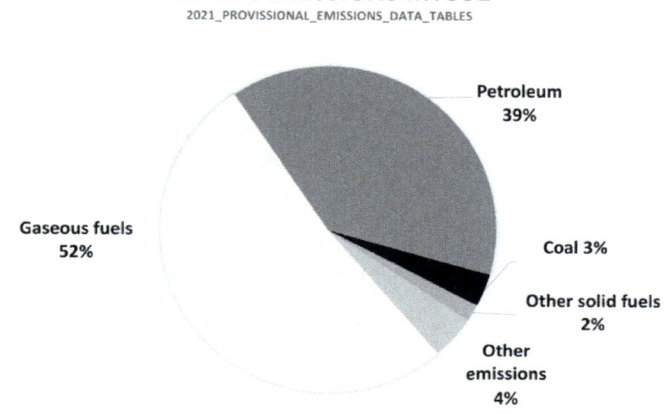

Fuel source for the UK's 2020 carbon dioxide emissions is predominantly from two sources - petroleum and natural gas.

Energy – End Use (2019) [BEIS Energy Consumption in the UK]

2.38 In 2019 UK's total energy used (all fuels and sources) was 142,000 ktoe (thousand tonnes of oil equivalent). Converting this to Watt hours gives ~ 1,654,000 GWh.

2.39 'Transport' is dominated by burning oil (petroleum) whilst 'domestic' (mostly heating and hot water) is dominated by burning natural gas.

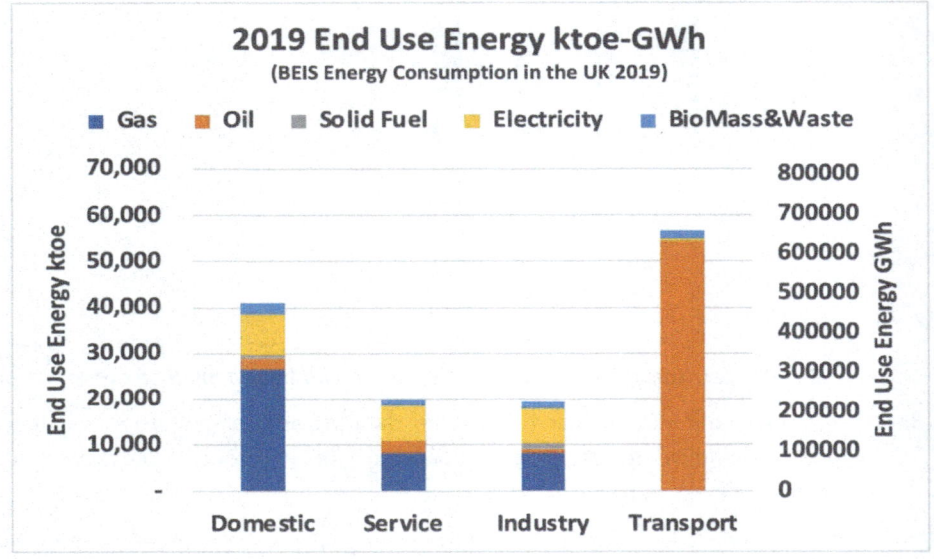

Chart shows the end use of fuels

Lifecycle greenhouse gas emissions

2.40 Whilst there is some variation in the amount of greenhouse gases per unit of energy produced by each fuel source over the lifetime of the generation, the trend is clear - fossil fuels are heavy emitters whilst renewables are light emitters

2.41 The following chart has followed the UNECE analysis in 2020 with some additions for biomass and oil from the World Nuclear Association. All analyses considered gave the same broad picture.

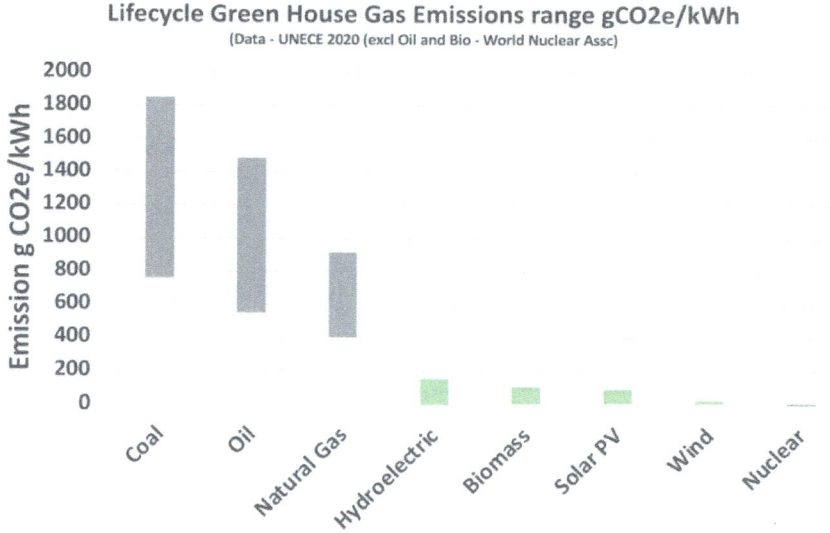

Fossil fuels are heavy emitters of greenhouse gases. Natural gas, often put forward as a clean transition fuel, is a significant emitter. Non-fossil fuel energy sources are considerably lower, including nuclear.

3. Toxic emissions

Combustion of any fuel, fossil or otherwise, leads to harmful toxic emissions.

Particulates formed from burning, including those that are not visible, are very harmful; the smaller the particulate the further into the body it will reach.

Any fuel, including hydrogen, when burnt in air will produce nitrogen dioxide (NOx).

Other combustion products include carbon monoxide, methane, and other hydrocarbons, depending on the fuel.

The UK's air quality guidelines fall short of those from the World Health Organisation (WHO) air quality guidelines.

A 'clean air zone' preventing the use of solid fuel burning (including wood burners) should be implemented in all major conurbations.

According to the World Health Organisation, Plymouth has worse air quality than London. In 2016, in a list of UK towns and cities, Plymouth was ranked sixteenth out of the thirty-two towns tested. [31][32]

Toxic Emissions

3.1 Toxic emissions arising from the combustion of fossil fuels affect air quality and consist of two types – gases and particulates.

3.2 Toxic gases include carbon monoxide (CO), methane (CH_4), non-methane hydrocarbons (NMHC), hydrocarbons (CH) and nitrogen oxides (NOx).

3.3 NOx is of particular concern and is a direct hazard to health responsible for aggravating cardiovascular and respiratory disease. NOx arises from combustion.

3.4 Toxic particulates can penetrate deep into the lungs and fine particles can cross from the lungs into the blood stream, including crossing to the foetal side of the placenta.

3.5 Particulates are categorised as:
- PM_{10} (coarse 10μm to 2.5μm): these will reach into the upper respiratory tract,
- $PM_{2.5}$ (fine 2.5μm to 1μm): these will reach into the lower respiratory tract,
- PM_1 (inhalable 1μm to 0.1μm): these will reach deep into the alveoli, in the lungs, where gas exchange takes place,

- $PM_{0.1}$ (ultra-fine <0.1µm): are the most hazardous and can reach deep into the bloodstream and the whole body, including crossing the placenta and entering the brain.

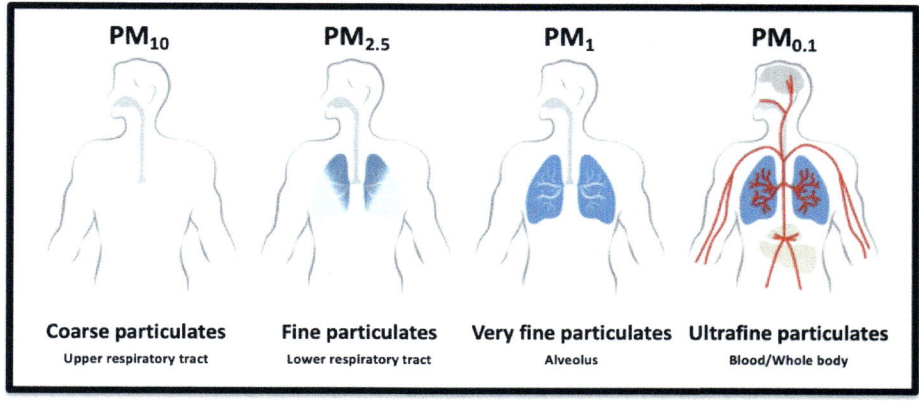

Penetration into the body of particulate combustion emissions [33] The smaller the particle, the further into the body it can penetrate. Very fine and ultrafine particles are not visible under normal circumstances but must not be ignored.

3.6 The finer the particle the more hazardous it is. "Exposure to particles can lead to mortality (death), increased admissions to hospital of people suffering from cardiovascular (heart) disease (attacks and strokes), and pulmonary (lung) disease, such as chronic obstructive pulmonary disease (COPD), bronchitis and asthma. Other compounds found on particulates, such as some hydrocarbons or metals, can cause cancer or poisoning" [34]. The most recent evidence from the Government's Committee on the Medical Effects of Air Pollution (COMEAP) [35] suggests that some 29,000 deaths per year are brought forward by exposure to man-made particulate air pollution at current levels.

Emission standards

3.7 The UK had a stated intention to adopt the more stringent World Health Organisation (WHO) particulate emission standards rather than the existing EU standards despite failing to even achieve those. [36] This means an annual mean exposure level for $PM_{2.5}$ of 5µmg/m³ down from 25 and for PM_{10} a limit of 15µmg/m³ down from 40.

3.8 Plymouth does not currently meet WHO levels of emission, albeit it does meet current UK/EU levels in most places except Exeter Street and Mutley Plain. [37] IQAir provides real time air quality measurements from three sites around the city; more are necessary. [38]

3.9 Disappointingly, Plymouth has no plans to restrict emissions from wood burning stoves.

4. The Energy Market, Fuel Taxation, Levies, Contracts for Difference

Fuel bills make up a disproportionate cost for the poorest in society.

Environmental and social levies are applied mostly to electricity bills. Electric heating is disproportionately used by the poorest in society, often in poorly insulated rental properties.

Network costs are applied to the daily standing charge and so are not usage dependant.

Costs associated with energy company failures are recovered in the network charge and so are disproportionately incurred by the poorest in society.

Although less than 50% of electricity is generated by natural gas, it is the wholesale price of gas that determines energy costs despite the substantial amount of renewable energy on the UK grid.

Energy Market

4.1 Gas and electricity supply involves generators/transporters, transmission, local distribution networks, and energy suppliers. [39]

4.2 Electricity generators provide power to the national grid or, if under 150MW, direct to the local distribution network. Gas is delivered to the National Transmission Network by the gas transporters.

4.3 Transmission/local distribution is two-tier. The National Grid transmits electricity at high voltage to eight local distribution networks from where it is distributed at lower voltage to the end user. The gas National Transmission Network transports gas at high pressure to eight local distribution networks which operate at lower pressure and deliver gas to the end user.

4.4 Energy suppliers purchase energy on the open market, often well in advance, and because of fuel price volatility this usually involves hedging. They provide billing, payments and meters to properties and administer government schemes such as the Energy Company Obligation, Feed-In tariffs, and Warm Home Discount. Energy suppliers are responsible for meeting the Renewable Obligation determining how much renewable energy is in their electricity mix.

Utility Bills

4.5 Electricity and Gas Utility Bills are made up of

- Wholesale costs
- Network costs (including 'supplier of last resort' costs for failed suppliers)
- Policy, environmental, and social costs
- Operating costs
- Earnings Before Interest and Taxes (EBIT)
- VAT @ 5%
- Direct debit and headroom

Environmental and social levy

4.6 When the grid was dominated by coal-generated electricity, loading electricity bills with environmental costs may well have had some logic – tax what you don't want and subsidise what you do.

4.7 Encouraging gas usage for heating may well have made sense then – it no longer does in 2020 when the grid is lower carbon than gas boilers. [40]

4.8 The recently released **British Energy Security Strategy (BESS)** sets out the intention to 'rebalance the policy levies and costs placed on energy bills away from electricity to incentivise electrification across the economy'. This is reinforced by the **Heat and Buildings Strategy** [41] which sees the shift of levies from electric to gas as being essential to ensure that heat pumps are no more costly to run than existing gas boilers.

4.9 The **BESS** also sets out to expand the **Energy Company Obligation (ECO)** to £1bn per year from 2022 to 2026. [42]

4.10 Whereas the rebalancing of policy levies from electricity to gas should be broadly neutral in terms of costs to consumers, the additional ECO, currently funded largely through electricity bills will disproportionately penalise poorer families.

Climate Action Plymouth – Sustainable Power and Energy Group

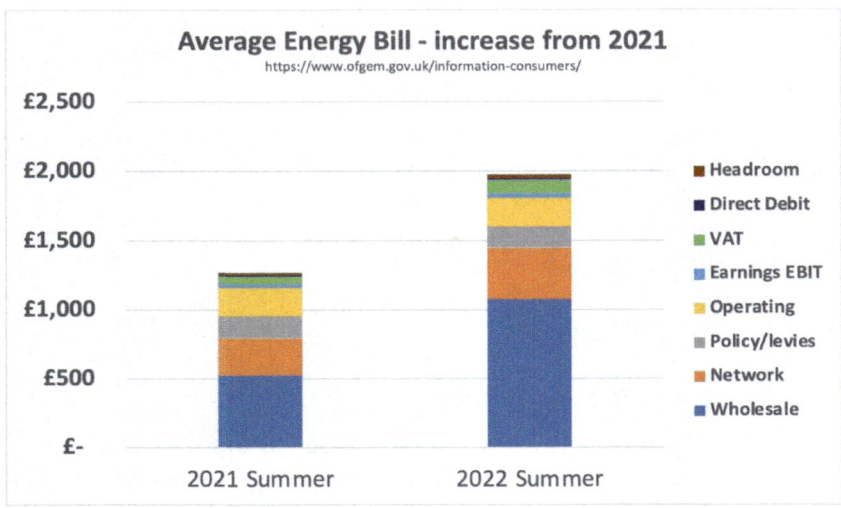

Typical dual fuel customer paying by direct debit. Volatility in the market due to gas prices is evident and is likely to continue
(Source Ofgem [43])

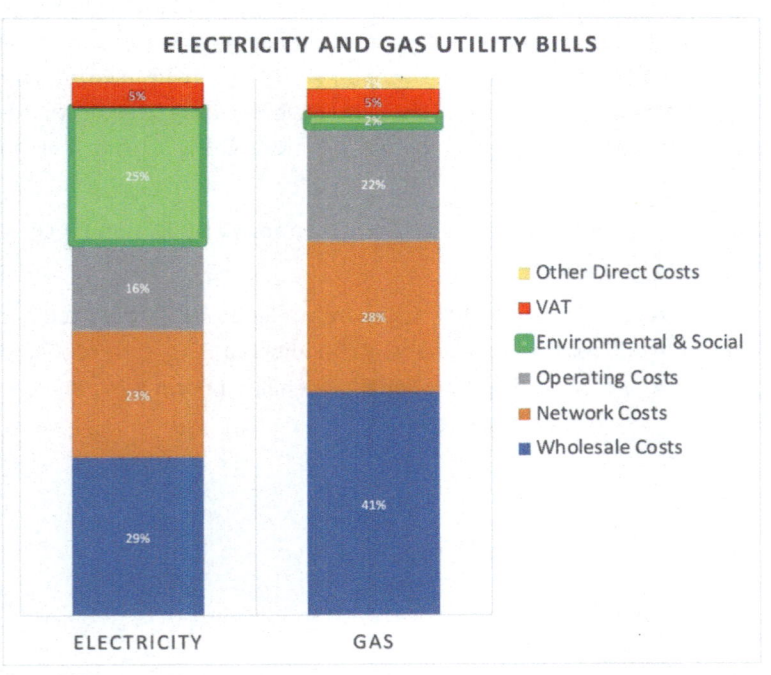

Make up of gas and electricity bills 2019 showing the disproportionate weight put on cleaner electricity compared to natural gas.

4.11 A breakdown of electricity and gas bills (2019) by OFGEM shows the disparity in the environmental and social levy between the two fuels - 25% of a typical electricity bill but only 2% of a gas bill!

4.12 Dwellings using electric (rather than natural gas) for heating and hot water tend to be of lower energy efficiency and households of lower income. [44]. Direct electric heating and hot water is associated with **fuel poverty** not because of the efficiency of the electric system but because of the much higher electricity unit energy cost than that of natural gas. Social and environmental levies disproportionately fall on those households with electric heating, exacerbating the fuel poverty issue.

Contracts for Difference

4.13 Before the 1990s, provision of electricity was a government-controlled monopoly. Since 1989, the government developed a 'fully competitive market' giving customers the freedom to choose their energy suppliers. Energy suppliers buy their electricity from electricity generators, those that produce electricity from power stations. This allows wholesale electricity transactions to take place in a commodity market on a day-to-day basis.

4.14 Low carbon electricity is protected from this commodity market by the 'Contracts for Difference' mechanism. "Successful developers of renewable projects enter into a private law contract with the Low Carbon Contracts Company (LCCC), a government-owned company. "The CfD is based on a difference between the market price and an agreed "strike price". If the strike price is higher than the market price, the CfD counterparty must pay the renewable generator the difference between the strike price and the market price. If the market price is higher than the agreed strike price, the renewable generator must pay back the CfD counterparty the difference between the market price and the strike price. [45]

4.15 The UK Government's Review of Electricity Market Arrangements [46] recognises that the wholesale gas price sets the price of electricity, although it contributes less that 50% of the fuel used to generate electricity, even where the cost of renewables is lower. 'Wholesale prices are set by the variable cost of the "marginal plant"; a system that can react quickly to electricity demand, such as a gas fuelled gas turbine.' [47]

4.16 Contracts for Difference are now an annual auction. [48]

5. Energy Use/Demand Reduction

Household demand for energy has reduced over the past decade but must reduce further by 2030 to tackle climate change.

All new housing stock should be built to high standards of insulation and ventilation.

All existing housing stock should have an energy efficiency improvement plan with incentives applied to encourage it to happen.

Payback times on insulation measures are now a third of the time or less than before the current energy crisis.

Ventilation must be considered when insulating older properties.

Reform 'Energy Performance Certificates' (EPC) to reflect energy usage rather than cost – this will keep EPCs valid after levies and taxes are redistributed from electricity to gas/fossil-fuels.

UK Government strategy and initiatives

5.1   The UK Government has three strategy documents each having different funded initiatives to support them.

5.2   The **British Energy Security Strategy (BESS)** aims to improve energy efficiency, upgrade 450,000 homes over a four-year period, and ensure that by 2025, all new buildings will be ready for net-zero. To support that aim, it has **cut VAT on insulation to 0%** for five years, introduced a **Boiler Upgrade Scheme** (2022) to encourage the shift to heat pumps [49], a **Green Heat Network Fund (GHNF)** to incentivise the wider adoption of low carbon heat **networks** [50], and a **Social Housing Decarbonisation Fund** [51].

5.3   The **Green Industrial Revolution (GIR)** was introduced to make the UK a global leader in green technologies when emerging from the global coronavirus pandemic. The **Homes Upgrade Grant is available** for off grid homes energy improvements [52]. **The implementation of the Future Homes Standard** [53] should ensure that all homes built by 2025 will produce 75-80% fewer carbon emissions than those built today (2022) and that homes built to that standard will require no upgrade to meet the 2050 net zero target. The GIR requires mandatory disclosure of the **Energy Performance Certificate**. The **Green Homes Grant** and **Public Sector Decarbonisation Scheme** have both been withdrawn.

5.4   The **Heat and Building Strategy (H&BS)** identifies £1.8bn for a **Home Upgrade Grant** (also identified in the GIR), a **Social Decarbonisation Fund** (also mentioned in the

BESS) and for commitments included in the **Future Homes Standard** (also included in the GIR). New commitments include reducing energy consumption of commercial and industrial buildings by 2030, public sector building emissions by 75%, supporting social housing, and low income and fuel poor households.

5.5 The **Help to Heat** scheme, announced in September 2022, provides a £1.5bn package to improve energy efficiency of UK social housing and low-income properties. [54]

5.6 A consultation **Delivering a smart and secure electricity system: the interoperability and cyber security of energy smart appliances and remote load control** [55]] seeking views on a smart and secure energy system has recently been launched. It will look at ways to decarbonise the power system by 2035, to support energy independence, and to achieve net zero by 2050. The consultation considers how to control large electrical loads (>300MW) as well as smaller domestic loads such as EV Charging, batteries, heat pumps, and heating appliances. Time-of-use tariff, control of smart heating appliances, and a flexible licensing framework for domestic users are all for consideration.

PCC's CEAP initiatives

5.7 Delivering energy efficiency improvements to over 150 homes by 2023, advising 200 private landlords on energy improvements

5.8 Gaining planning permission for Energiesprong low carbon housing [56]. Develop site infrastructure at Kings Tamerton and identifying further potential developments, both new and retrofit, for Energiesprong principles working with Plymouth Community Homes and Live West.

5.9 Interrogating **Future Homes Standards** and **Future Buildings Standards**.

5.10 Implementing web-based home assessment tool together with Plymouth Energy Community (PEC)

Home energy use

5.11 Homes use energy for space heating/cooling, hot water, appliances (dishwasher, washing machine, TV, anything plugged in, etc), lighting, and cooking. Heating and hot water are the dominant usage and are usually powered by natural gas. Appliances are the dominant electrical usage.

5.12 Reducing energy demand is essential to tackling the climate crisis and will be critical to enabling the move from gas heating to electric heat pumps.

5.13 Solar panels have accounted for a reduction in grid electricity used (not overall energy usage) of about 14% [NEEDS Framework 2021].

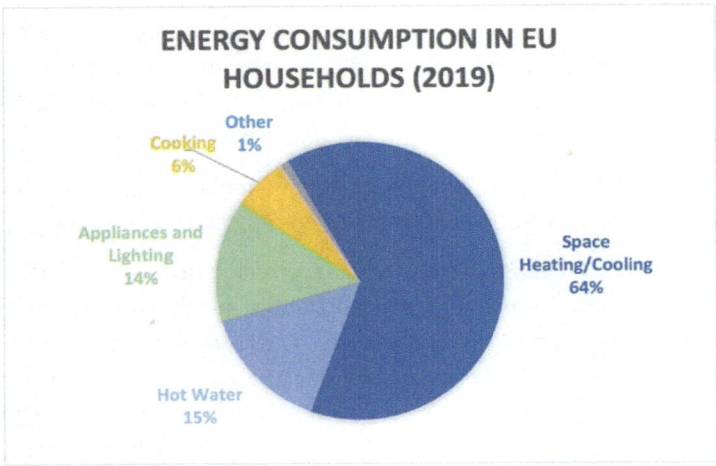

*Typical final energy usage in an EU household [57].
The typical UK user is expected to be similar.*

Space heating

5.14 Space **heating and hot water** are mostly provided by fossil fuel – 85% of households are connected to the gas grid, the remainder use oil, LPG or wood with some households using direct electricity. There is a slowly increasing number using electric heat pumps.

5.15 Reducing the energy used for space heating of homes involves **increasing loft Insulation** ~4% and **adding wall Insulation** (solid wall ~18%, cavity wall ~9%). Insulating some of our pre-war housing stock is very challenging.

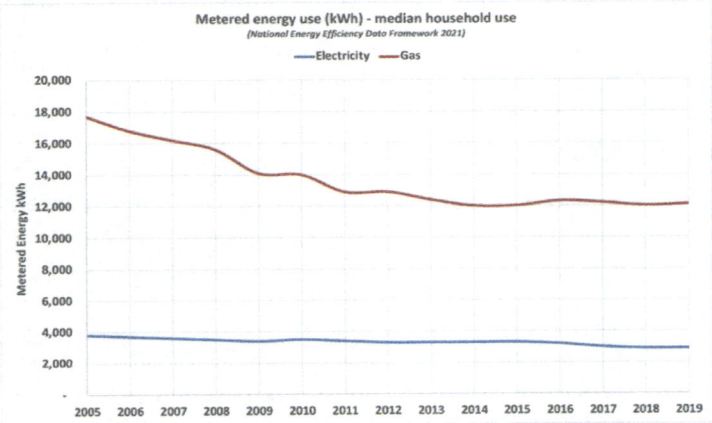

Metered household energy demand has decreased since 2005 for both gas and electricity by about a third in each case. This reduction has come from condensing boilers, from increased insulation, and from improved efficiency of appliances.

5.16 An energy reduction strategy of replacing conventional boilers with more efficient **condensing gas boilers** has reduced energy used by `5%, although often the setting up, optimising, and balancing of the condensing boiler system has left a lot to be desired

Domestic electrical usage

5.17 **Appliances and lighting** are powered by electricity. [58] 'Wet' refers to washing machines and dishwashers. 'Cold' to refrigerators and freezers.

5.18

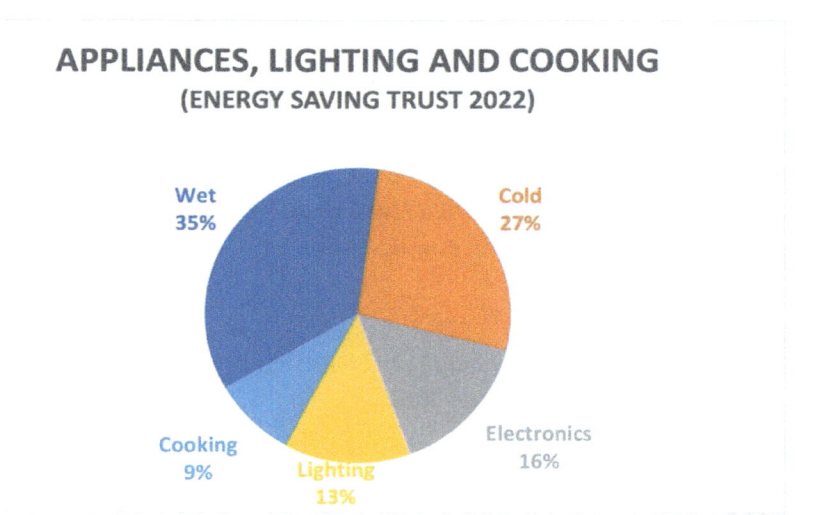

APPLIANCES, LIGHTING AND COOKING
(ENERGY SAVING TRUST 2022)

- Wet 35%
- Cold 27%
- Electronics 16%
- Lighting 13%
- Cooking 9%

Chart of the typical energy split between 'wet' (dishwashers, washing machines, etc) and 'cold' (refrigerators and freezers), cooking (hob and oven), lighting, and electronic appliances (TVs, computers, etc.)

5.19 More efficient appliances (particularly cold appliances), the change to low energy lighting, and the improvement in efficiency of computers, TVs and other electronics have also contributed to reduced household electricity usage. [59]

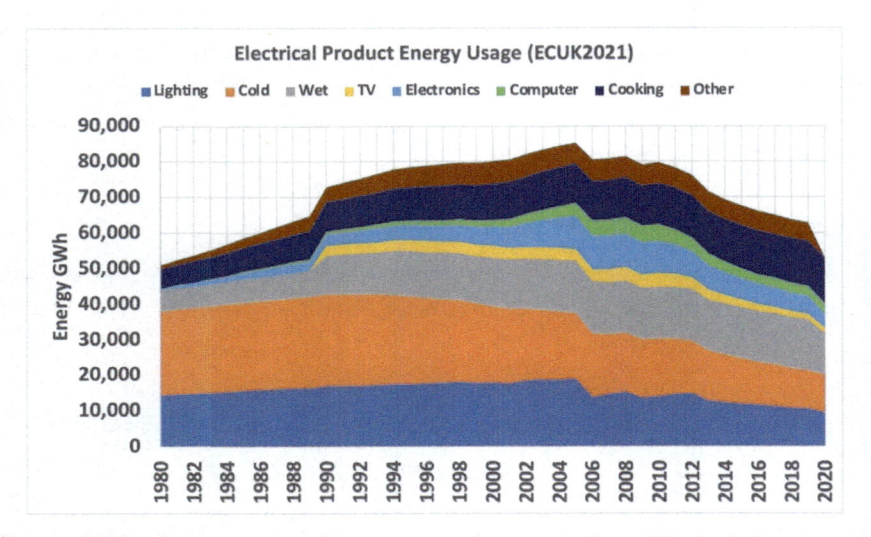

Climate Action Plymouth – Sustainable Power and Energy Group

Electrical Product Energy Usage (ECUK2021)

Regulations governing the efficiency of electrical appliances have led to energy demand returning to 1980s levels when the amount of electrical goods were significantly fewer and computers and other electronics were yet to be seen.

England's housing stock

5.20 The following energy standards are at a national level. There is no reason to believe that things are much different locally.

5.21 Reducing energy demand sufficiently to ensure that heating can be provided by zero-emission electricity, including heat pumps, will require improvements in energy efficiency across our housing stock.

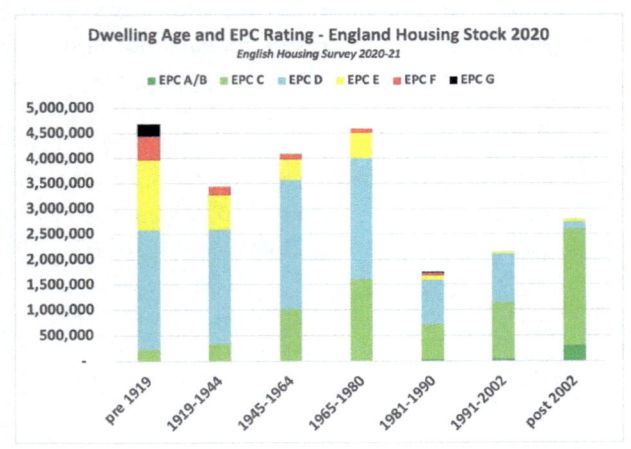

Dwelling Age and EPC Rating - England Housing Stock 2020 (English Housing Survey 2020-21)

England's housing stock is dominated by pre 1980s building standards. Note: Up to 1930, typically solid walls; unfilled cavity walls from 1930-1995; filled cavities from 1995 onwards.

ENGLAND HOUSING STOCK 2020 - EPC RATING
ENGLAND HOUSING SURVEY 2020-2021

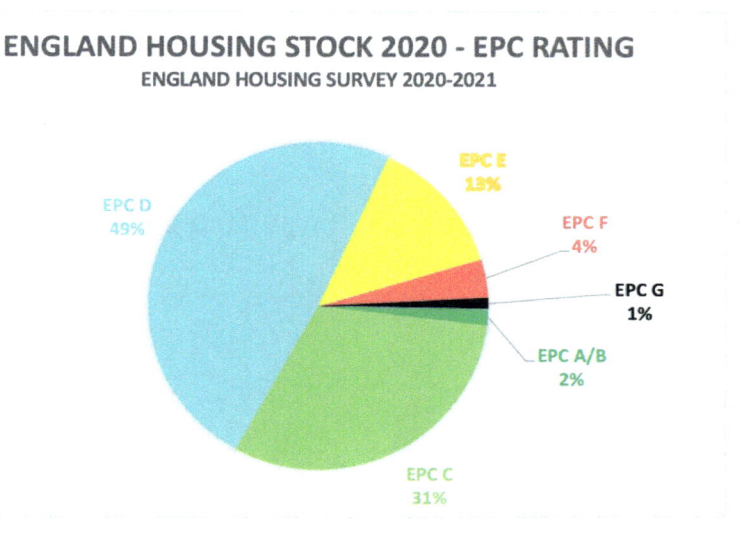

- EPC D 49%
- EPC E 13%
- EPC F 4%
- EPC G 1%
- EPC A/B 2%
- EPC C 31%

RIBA Architectures 'Homes for Heroes' report on Inter-War Housing suggests that retrofitting all interwar houses to EPC of C or better would reduce energy by 15GWh and save 8.5mtCO$_2$e annually. [60]

Insulation measures, including double glazing.

Insulation Measures 2010-2020
English Housing Survey 2020-21

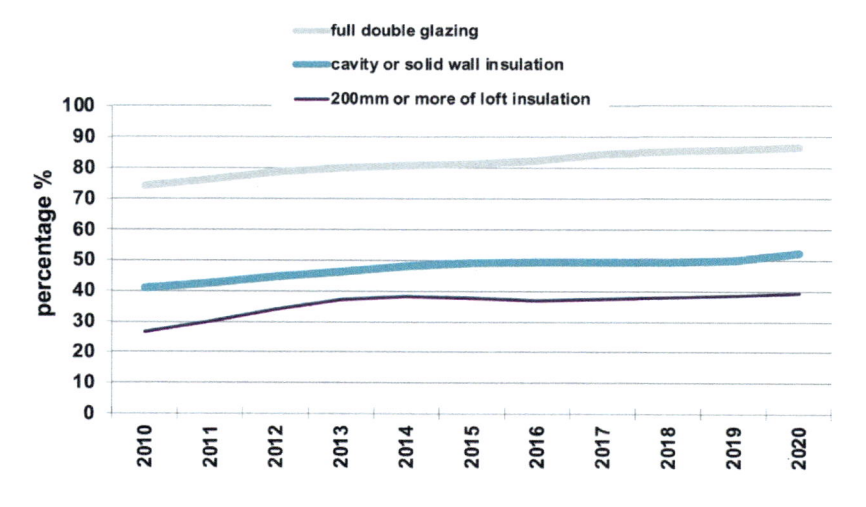

- full double glazing
- cavity or solid wall insulation
- 200mm or more of loft insulation

Energy improvements undertaken to homes has included double or triple glazing (almost 90%); increasing loft insulation (just 40%); and Insulation to external walls (around 50%).

5.22 **Double glazing** is approaching 90% of current housing stock. Significant improvements to the performance of glazing can be achieved by replacing single glazing and by updating older double glazing with the latest standards. Whether triple glazing is effective in updating older properties is questionable but should be mandated in all new builds.

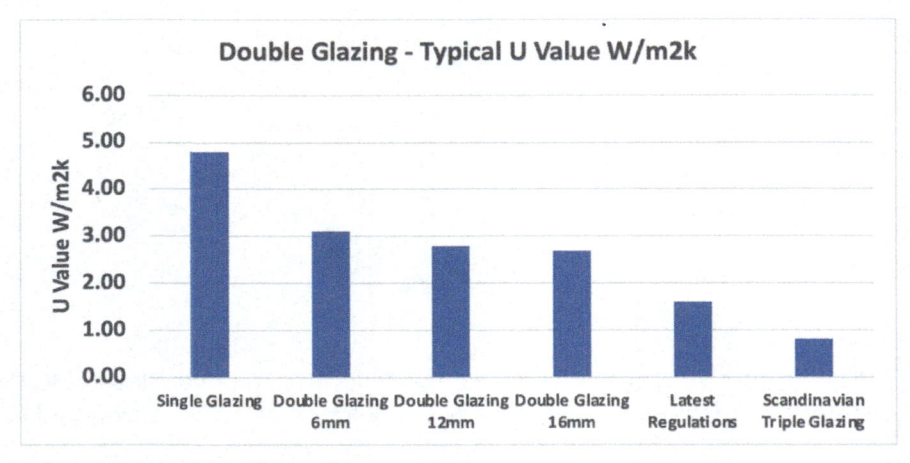

Very high insulation performance can be achieved by modern triple-glazing

5.23 **Loft insulation** can be done several different ways: blanket, loose fill, spray foam, insulation boards, and insulation slabs. They each have their benefits and drawbacks.

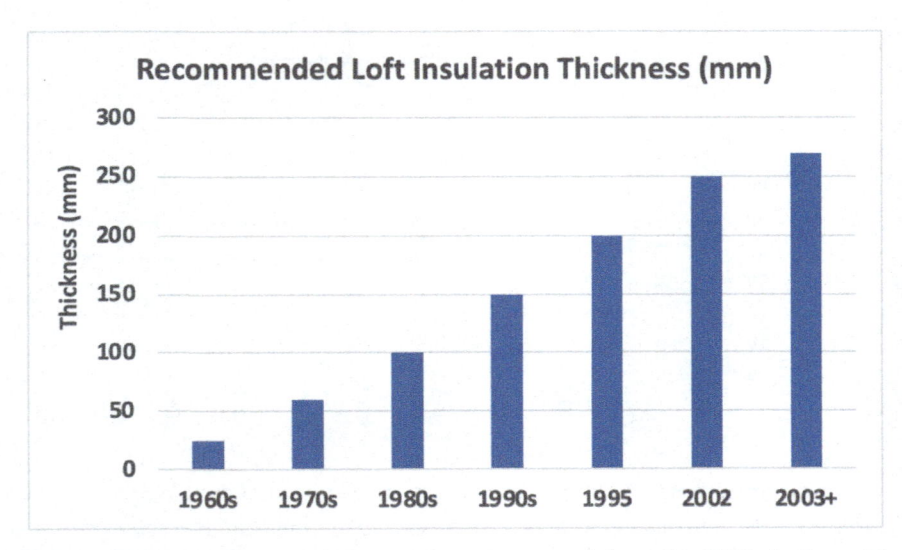

Loft insulation thickness regulations have increased since the 1960s in line with improving home insulation.

5.24 An uninsulated loft space has a U-value of 2.5W/m²k. 200mm of rockwool provides a reduction to 0.2W/m²k. Fewer than 40% of England's housing stock has insulation at this level. A more expensive multi-foil system could deliver insulation at 0.04W/m²k or lower.

5.25 Great care should be taken in a loft to ensure adequate ventilation/airflow in the loft space to avoid condensation, dampness, and mould. Increasing loft insulation makes the loft colder and more prone to damp and condensation.

5.26 Applying insulation has the effect of reducing heat loss from inside the house but it also has the effect of reducing heat intake during the summer months – the house will maintain a more even temperature all year round.

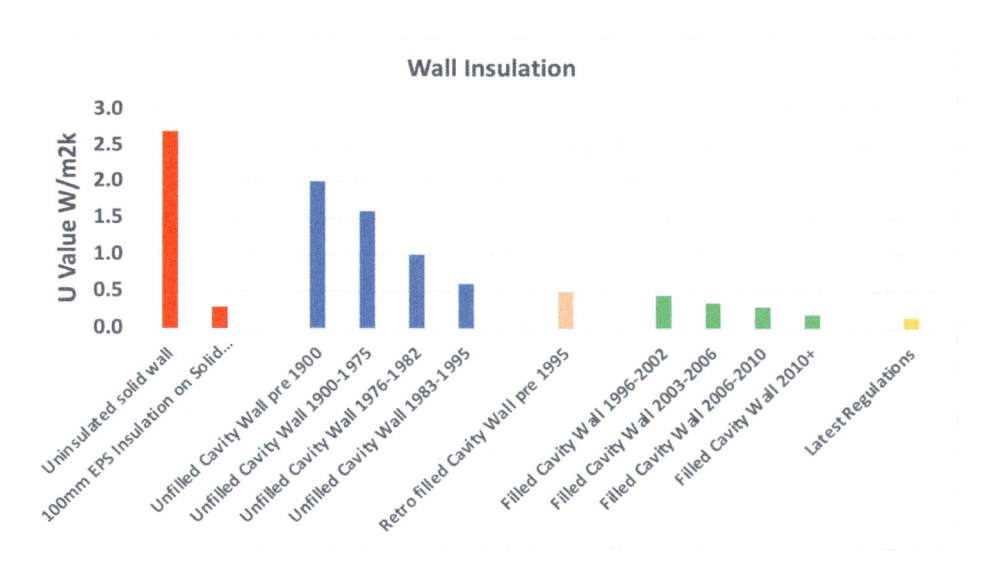

Typical insulation values for walls

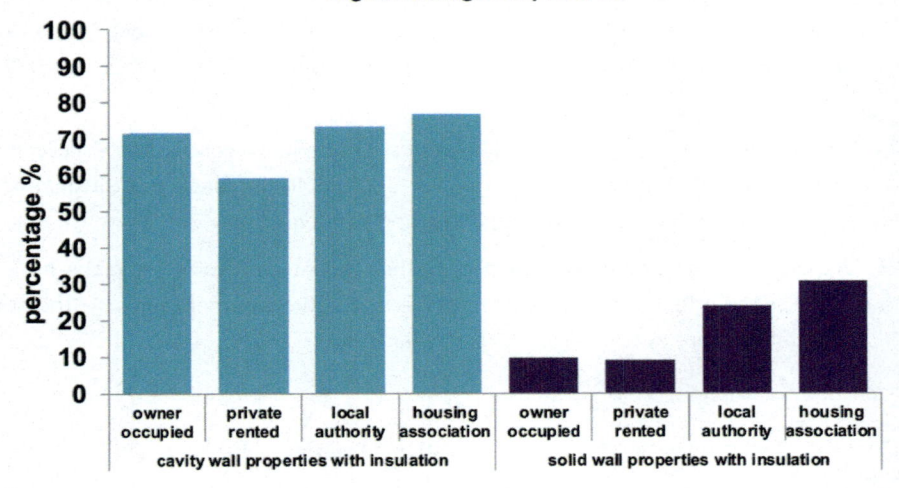

Wall insulation, by main wall type and tenure 2020
English Housing Survey 2020-21

Housing Associations and Local Authorities have made most progress in applying wall insulation to older properties. Cavity wall insulation has a good market coverage, but more can be achieved.

Ventilation

5.27 Once insulation has been installed, draughts can become the biggest heat loss. However, older properties require ventilation to avoid damp and condensation issues.

5.28 Heat recovery ventilation systems are available, extracting moist air, passing it through a heat exchanger and warming incoming air. Systems can recover up to 95% of the heat energy whilst ensuring clean fresh dry air. [61]

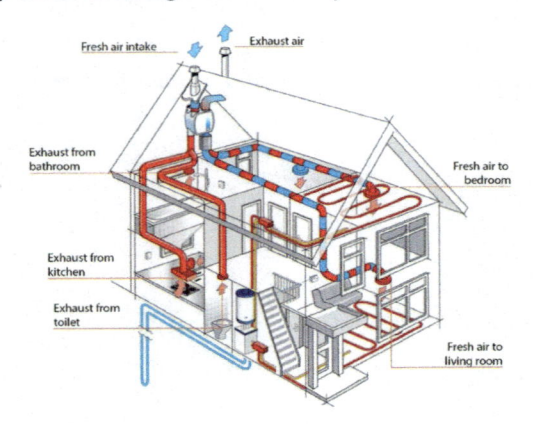

Mechanical Heat Recovery Ventilation System *(illustration Customradiant.com)*

Passivhaus

5.29 "The Passivhaus standard is designed to create buildings that are both comfortable and energy efficient. The combination of an air-tight, super-insulated building fabric and mechanical ventilation with heat recovery reduces energy requirements by up to 85% and provides a draught-free environment with excellent air quality." UK Passivhaus Conference 2010. [62]

5.30 Primrose Park in Plymouth is the largest residential Passivhaus development in the UK. The development by Plymouth Community Homes was in partnership with Plymouth City Council. [63]

Houses in Primrose Park, Plymouth, built to Passivhaus standards (Photo Malcolm Teague)

Industrialising building insulation – Energiesprong

5.31 Dutch Company Energiesprong [64 65], have industrialised the building insulation process, shifting much of the work into the factory and minimising on-site work. Whether such a process could work with UK housing stock, which is more varied than Dutch building, is yet to be seen but is a pointer to what is necessary to bring retrofit insulation costs down.

Retrofitting a Victorian terraced house

5.32 The Aston Group retrofitted a traditional (1902) Victorian terraced house in Walthamstow with the aim of changing its EPC from a band E to band A. [66] The budget for doing the principal energy saving features was £37,000 out of a total budget of £112,000 for the refurbishment. The project was undertaken between January and June 2021. Whether such a scheme demonstrates value for money depends upon the prevailing price of houses and the cost of energy but technically such retrofitting is certainly feasible.

5.33 The retrofitting included:

- External Wall Insulation (EWI) (90mm EPS rendered) to side and rear but Internal Wall Insulation (IWI) to the front (57.5mm insulated plasterboard).
- Roof insulation upgraded to 300mm of mineral wool.
- Underfloor insulation using sprayed foam applied by Q-Bot.
- Replacement of gas boiler with Air Source Heat Pump (ASHP), hot water cylinder and radiators.
- Photo Voltaic (PV) array (3.9kWp) together with battery storage.
- High performance double glazing
- Mechanical ventilation to improve air quality and provide fresh air.
- Waste-water heat recovery system

5.34 Walthamstow – Costs and Payback

	Cost excl VAT	CO2 saving g/pa	CO2 saving g/pa per £ investment	Cost Saving (2021)	Simple Payback yrs	Cost Saving (2022)	Simple Payback yrs
Loft Insulation	£ 829	580	0.700	£ 120	7	£ 360	2
ASHP and Radiators	£ 10,200	4240	0.416	£ 860	12	£ 2,580	4
Underfloor Insulation	£ 1,850	287	0.155	£ 67	28	£ 201	9
Solid Wall Insulation	£ 12,081	890	0.074	£ 341	35	£ 1,023	12
Solar PV and Battery Storage	£ 12,294	853	0.069	£ 609	20	£ 1,827	7
Total	£ 37,254	6,850		£ 1,997	19	£ 5,991	6
Total Refurbishment Project	£ 112,000						

Payback duration is greatly affected by the extant price of heating and electricity. 2022 has seen a very large rise in energy prices and hence any savings made will shorten the investment payback period.

5.35 In cost terms, investment payback time and CO_2 reduced per investment, loft insulation is the most attractive followed by the change of a gas boiler to ASHP. Underfloor, solid wall insulation, and PV solar are less attractive on one or more of the criteria.

Plymouth's housing stock

5.36 England Housing 23.5m dwellings. Plymouth has 112,477. Plymouth has a broadly similar housing mix by tenure as nationally.

5.37 England's interwar housing stock accounts for 3.3m homes, most of which have solid wall construction. Retrofitting could save up to 4% of England's total CO_2 emissions. [67]

5.38 Plymouth Housing Plan 2012-2019 [68]

5.39 In comparison with the UK and much of the Southwest, Plymouth has relatively:

- Lower levels of owner occupation and higher levels of private rented and social rented homes.
- More, older, privately owned or rented properties in poor standard or failing basic tests.
- Higher concentrations of poorly maintained and fuel inefficient private housing occupied by older and vulnerable, low-income households
- More poor standard social rented homes needing renewal and regeneration in mono tenure estates.

5.40 It is estimated that there were nearly 14,000 households living in fuel poverty in 2010. Many of these people live in older homes and face a long-term fuel-poor future with a high probability of poor health unless we can either improve hard-to-treat homes or enable fuel poor householders to move into more energy efficient housing. [Plymouth Housing Plan 2009-2017) [69]

Plymouth's progress with carbon saving on municipal buildings

5.41 The following measures are being installed on Plymouth City Council buildings from 2021 to 2022 [70]

- Heat pumps   402t CO_2/year
- Insulation works   11t CO_2/year
- Secondary glazing   4t CO_2/year
- Solar PV   188t CO_2/year
- Lighting (LEDs)   160t CO_2/year

ENGLAND - DWELLING BY TENURE 2020
ENGLISH HOUSING SURVEY

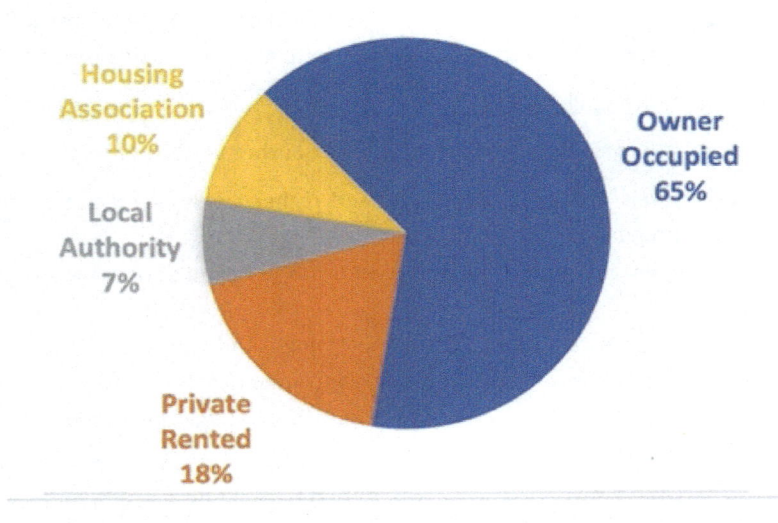

- Owner Occupied 65%
- Private Rented 18%
- Housing Association 10%
- Local Authority 7%

PLYMOUTH - DWELLING BY TENURE 2009-2017
PLYMOUTH CITY COUNCIL

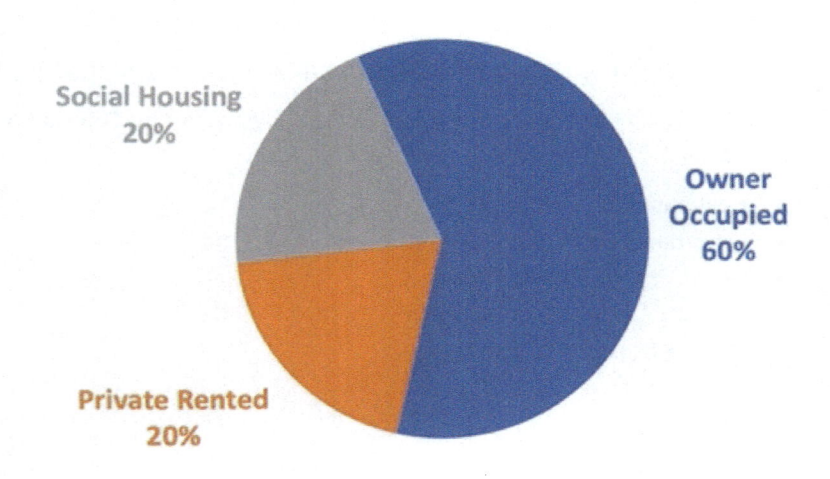

- Owner Occupied 60%
- Social Housing 20%
- Private Rented 20%

6. Heating and hot water

Heating and hot water is predominantly achieved by burning natural gas.

The move to condensing gas boilers in place of conventional gas boilers has only been partially successful due to poor installations. Too many are set with too high a flow temperature meaning the boiler is not able to condense the flue gases and hence efficiency is lost. Because of the lower flow temperature, larger radiators should have been considered for retrofits, but this is rarely, if ever, done.

To achieve a high efficiency, a condensing boiler requires a lower optimum water temperature, weather compensation, load compensation, flue gas heat recovery and smart controls.

All domestic heating and hot water should be achieved either by direct electricity or by heat pump, wherever feasible. A high standard of training is required for a successful, efficient, heat installation. A much greater time is required to conduct the heat balance prior to the installation – customers are often unwilling to pay for this.

Where heat pumps are fitted, they should use refrigerants with a low GWP – 1 or below.

UK Government strategy and initiatives

6.1 The UK Government has three strategy documents and different funded initiatives to support them.

6.2 The **Green Industrial Revolution (GIR)** identifies the intention to fit 600,000 heat pumps per year by 2028. *(This is approximately 10% of England's total housing stock).*

6.3 The **Heat and Building Strategy (H&BS)** states the intention to increase the biomethane mix in the current gas grid, to end fossil fuel heating for dwellings not connected to the grid by 2026, and to phase out the installation of new natural gas boilers from 2035. *(This should lead to the end of most gas fuelled heating and hot water by 2050 as boilers reach the end of their useful life.)*

6.4 The **Heat and Building Strategy (H&BS)** focusses strongly on the widespread introduction of heat pumps. The intention is to provide financial support to meet their capital costs, make sure that by 2030 heat pumps are no more expensive to buy than gas boilers, grow the manufacturing supply chain to ensure that they are smaller, easier to install and use, and better to look at, and to shift energy levies from electric to gas to ensure that they are no more expensive to run than gas boilers. There is also the intention to ensure that the electricity supply can meet the extra demand and that the pumps can be connected quickly and affordably.

PCC's CEAP initiatives

6.5 PCC is rolling out ground source heat pumps [71] across the education and student facilities and investigating the feasibility of marine source heat pumps. A digital resource to assist households to assess the viability of heat pump retrofits for a range of typical Plymouth housing is to be rolled out in 2022.

Heating and hot water

6.6 Heating and hot water is currently mostly (92%) derived from burning hydrocarbons / fossil fuels. Only 8% of heating and hot water is derived from electricity. Hydrocarbons including bioenergy/waste are used to produce approximately half of electricity generated (2020).

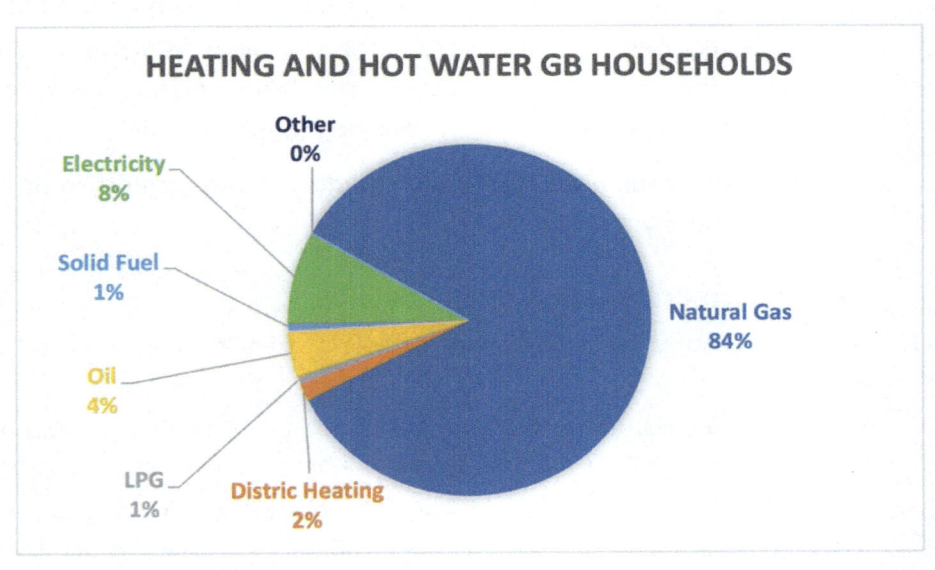

HEATING AND HOT WATER GB HOUSEHOLDS

- Electricity 8%
- Other 0%
- Solid Fuel 1%
- Oil 4%
- LPG 1%
- Distric Heating 2%
- Natural Gas 84%

Space heating and hot water are delivered predominantly from fossil fuels

End Use Heating and Hot Water

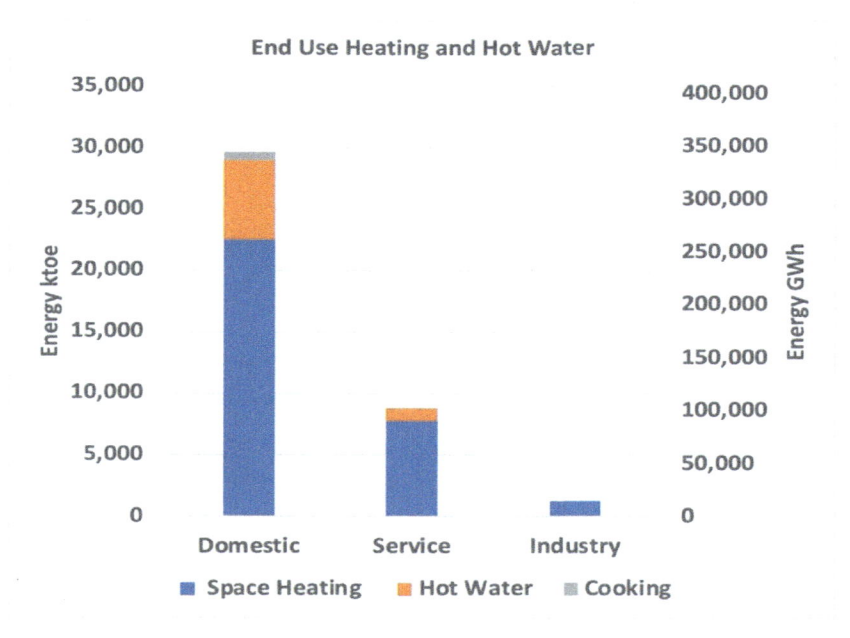

The energy use for space heating, hot water, and cooking (approx. 450,000GWh) is mostly used in domestic dwellings

Domestic Boilers

6.7 Domestic boilers can be arranged in a 'system', where the boiler delivers heating and hot water into a tank, or 'combi', where the boiler delivers heating with water heated on demand. Combi boilers are used more often on all dwellings rather than just for smaller homes and for flats. Combi boilers require a greater heating capability and where used for small amounts of hot water can be quite inefficient.

Gas/oil/solid fuel boiler

6.8 In urban settings, boilers are mostly fuelled by natural gas. In rural dwellings which are 'off-grid', oil, LPG, or solid fuel are used instead.

6.9 Domestic gas boilers have improved in their theoretical efficiency over the past decades. The greatest theoretical increase in efficiency came from the mandatory introduction of 'condensing' gas boilers from 2005 and condensing oil boilers from 2007.

6.10 A condensing boiler uses the direct gas flame to heat the water and uses the latent heat from condensing the flue gases to pre-heat the incoming cold water thereby significantly improving the boiler efficiency (around 10-15% improvement). When using the flue gases to pre-heat the incoming water, the gases will (or should if the water input or return temperature is around 45°C) 'condense' from a gas at about 200°C into a gas below 100°C together with the liquid condensate. The condensate (about 2-3 litres an hour) is an acidic waste product and drains into a waste-water outlet.

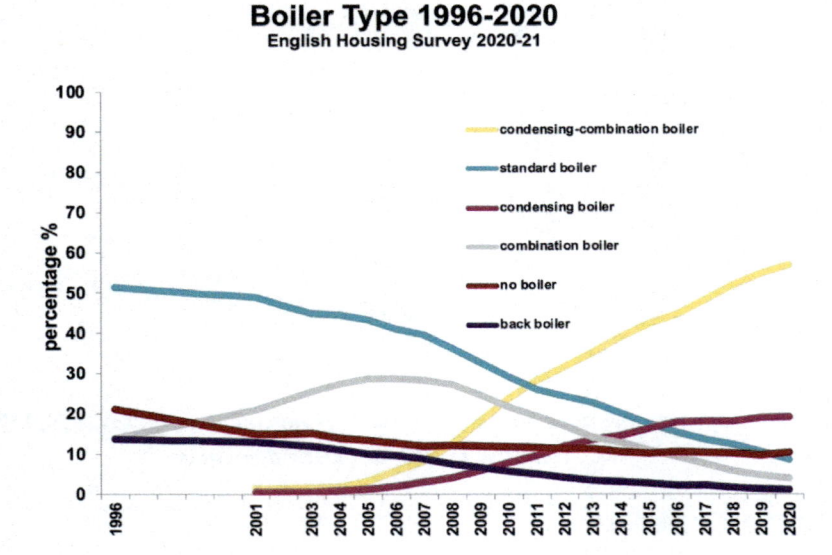

Boiler Type 1996-2020
English Housing Survey 2020-21

Condensing-combi gas boilers are now the dominant heating and hot water choice

6.11 A non-condensing boiler is usually set up to deliver heat to the radiators at 80°C and return it to the boiler at 60°C delivering the lost heat to the room. Older non-condensing boiler have an efficiency of around 55-70%. [72 73]

6.12 A condensing boiler, if it is to run in condensing (high efficiency) mode achieving 90%+, should have a return temperature of around 45°C meaning the temperature into the radiator should be around 65°C, about 15°C lower than a non-condensing, conventional boiler. At 55°C return temperature the efficiency drops to around 87%. At 60 °C return temperature, the boiler does not condense and so no efficiency gain is made! [74]

6.13 New central heating systems should have radiators capable of operating at the lower temperatures. For retrofit systems, where a non-condensing boiler is replaced with a condensing boiler, it seems unlikely that the radiators have been re-sized and so the system is likely to run inefficiently. The Energy Saving Trust found that condensing boilers typically deliver an efficiency nearer 80% than the 90%+ expected of them. [75]

6.14 To achieve an efficiency at SEDBUK ErP rating of 92% requires a lower optimum water temperature, weather compensation, load compensation, flue gas heat recovery and smart controls. This is not the case with most installations even in 2022.

6.15 The Carbon Intensity of a domestic gas boiler therefore varies significantly depending on the age and type of the boiler as well as how it has been set up. Boiler efficiency declines with age, even with regular servicing.

6.16 The Heating Hub suggests an 8% reduction in gas use by reducing flow temperature, but you may be on your own when it comes to resetting the boiler as they say 99% of boiler installers do not understand how condensing boilers work and therefore cannot set them up to run efficiently; this issue goes all the way back to 2005 when condensing boilers became mandatory. [76]

Electric boilers

6.17 Electricity is used for heating and hot water by just 8% of households in England, 5% in Wales and 13% in Scotland. Electric heating is more common in apartments (25%) with just 4% in houses. [Ofgem 2016 report]

6.18 Direct electric boilers can have an efficiency as high as 99%. Using off peak electricity and a well-insulated water tank can significantly reduce the carbon footprint. There are no grants to change from gas boilers to electric because electricity is still considered to be a high carbon source of energy. Whilst this may have been true in 2010 it certainly is not today (2022). Direct electricity has a carbon intensity lower than pipeline gas and much lower than LNG.

6.19 Direct electric heating and hot water is associated with 'fuel poverty' not because of the efficiency of the system but because of the generally lower efficiency housing [Ofgem 2016 report] and the much higher electricity cost than that of natural gas. Direct electric boiler heating is around 3 to 4 times more expensive than natural gas depending on the gas boiler efficiency (2020 fuel costs) or 2 to 3 times in 2022.

6.20 Social and environmental levies disproportionately apply to electricity and hence disproportionately fall to those households with electric heating, further exacerbating the fuel poverty issue.

6.21 Within Europe, where direct electric and heat pumps have a greater adoption rate than the UK, the important factors in the selection of electric heating is a fuel price ratio (electricity vs gas) in favour of electricity and the banning of fossil fuel boilers. Stringent building/insulation standards, a low carbon electricity supply grid, and the need for cooling are also important. [77]

Heat Pumps

Principles of a heat pump

6.22 Heat pumps use the refrigerant cycle to deliver either heating or cooling. The refrigeration cycle takes advantage of the change of state of a refrigerant from a liquid to a gas to cool or a gas to a liquid to heat. It is this change of state that enables a heat pump to deliver more energy than it uses.

6.23 Think of a heat pump as a refrigerator inside out - the grille on the back of the fridge is the condenser and gets hot as the refrigerant turns from a gas to a liquid and heat is rejected from it, whereas the inside of the fridge gets cold as the refrigerant evaporates from a liquid to a gas. The house is the outside of the fridge, and the inside of the fridge is the atmosphere/ground/water.

6.24 The refrigerant system uses:
- a compressor, to add energy to the refrigerant making it a higher-pressure gas at a hotter temperature
- a condenser, to turn the higher-pressure gas to a high-pressure liquid thereby releasing heat
- an expansion valve, to turn the high-pressure liquid to a low-pressure liquid, and
- an evaporator to turn the liquid to a gas, thereby extracting heat from the air/ground/water.

6.25 The refrigerant cycle can be from one to five times as efficient to use as direct electric heating – this is called the Coefficient of Performance (CoP).

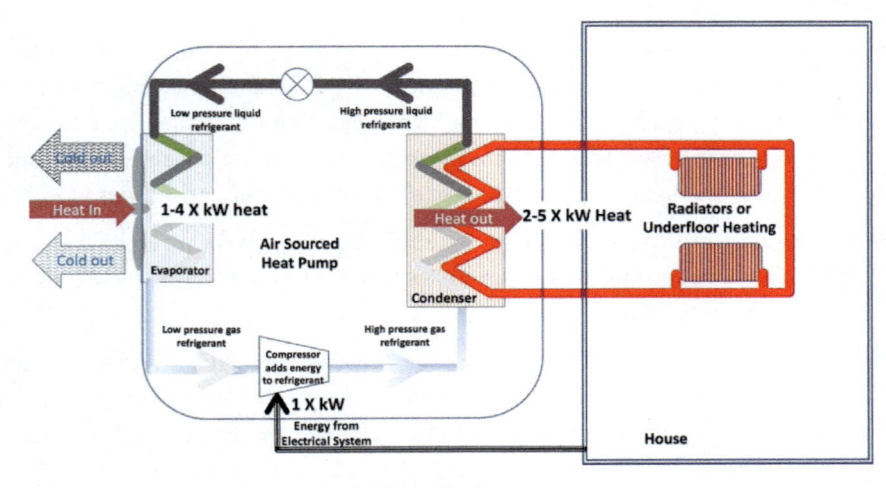

Principles of a heat pump

6.26 The compressor uses electricity and hence has the carbon intensity of the grid at that time divided by the efficiency (Coefficient of Performance CoP). The heat pump is therefore low carbon heating on today's grid-electricity. As the grid moves to zero carbon so the heat pump also becomes zero carbon.

Air-source/ground-source/-water-source

6.27 An air-source heat pump (ASHP) uses the atmosphere as the heat source. A fan ensures a large volume of air passes over the evaporator to enable the heat to be extracted. During very cold weather, when the heating is working hardest, the efficiency is lowest. An ASHP must be installed to allow good air flow and should not be within a metre of a neighbour's boundary. ASHP will become noisier as the ambient air temperature reduces and the compressor is made to work harder. The ambient night-time noise levels should be carefully considered, particularly in quieter suburban and rural areas.

6.28 A ground-source heat pump (GSHP) uses the ground as the heat source. The temperature of the ground below the surface is much more consistent and hence efficiency of the heat pump is similar across the seasons. Either a large garden will be required if the ground source is to be via trenches, or boreholes of some 75-200m deep. [78] As an example. a 107m borehole for a GSHP can be completed typically within a day.

6.29 A water-source heat pump (WSHP) uses a river, lake, or the sea as the heat source. The seasonal efficiency will be somewhere between the air and ground source HPs.

6.30 For new construction houses, a GSHP should be the preferred configuration. Wherever possible, a GSHP should also be used for retrofits with ASHP being fitted only where GSHP is not feasible.

Heat pump - central heating and hot water

6.31 A conventional boiler system is usually designed for a water temperature of about 80°C and a well set up condensing boiler system about 65°C. Heat pumps produce the highest efficiency when delivering a heating water temperature of between 35-45°C. [79]
With the much lower heating water temperature, larger bore piping and radiators that have a much larger surface area or, preferably, an underfloor heating system, will provide the best results. When using a lower heating water temperature, raising the temperature of the home can take a long time and so it is often preferable to run the heating system continuously.

6.32 A relatively recent product to market is the 'high-temperature' heat pump [80 81] offering a direct replacement for an existing gas boiler. Higher output temperatures are achievable either through a change in the refrigerant (e.g. to R744 CO_2 [82]) or using a two-stage cascade compressor. The capital costs of the high temperature HP if there is just a refrigerant change is likely to be similar or only marginally more expensive. The efficiency will be lower than the lower temperature HP but because existing piping and radiators can be utilised, overall costs can be lower. Wherever possible, low-temperature heat pumps are to be preferred.

6.33 A recent Electrification of Heat (EoH) project by BEIS found that there were no unsuitable homes for heat pumps. [83] Retro-fitting pre-1945 homes (and some homes in the 1945-1980 age bracket) with heat pump should be approached with extra care if efficiency and comfort are not to be compromised.

6.34 Whilst the new high-temperature heat pumps can deliver hot water on demand, [84] lower temperature heat pumps cannot, so there will be a requirement for a special larger hot water storage tank with larger heating coils storing hot water at a lower temperature. A typical bath water temperature is 40-45°C, so the tank should be sized to enable storage of all the bath water. The tank also boosts water temperature to 65°C at least once a day to avoid legionella.

Refrigerants

6.35 The refrigerants used most often for heat pumps are R410A, R32, R744, R1234yf, and R134A. Refrigerants can have significant Global Warming Potential (GWP) and Ozone Depletion Potential (ODP).

6.36 R-1234yf has a Global Warming Potential GWP of less than 1 [85], whereas R134a (an HFC) has a GWP of 1440, and R-12 (a CFC that is now no longer used) had a GWP of 8500. Some manufacturers [86] are adopting CO_2 (R-744) with a GWP of 1 - this comes with much higher pressures and other engineering challenges but enables much higher water output temperatures.

	GWP		Boiling Point (atm pressure) °C
Most used			
R410A (Pentafluoroethane)	2,090	Most ASHP and GSHP	-48.5
R134a (Tetrafluoropropene)	1550	Common for high temperature systems	-26.0
R513A (Hydrofluoroolefin) HFO	531	Drop-in replacement for R134A	-29.2
R32 (Difluoromethane)	675	Many ASHP from 2020	-52.0
Low GWP			
R290 (Propane)	3.3	Used in smaller systems	-42.2
R600 (Isobutane)	3.3	Used in smaller systems	-11.7
R744 (CO2)	1	Better for higher temperature output.	-78.0
R1234yf (Tetrafluoropropene)	<1		-29.0
R717 (Ammonia)	0	Potential future refrigerant	-2.2

Coefficient of Performance (CoP)

6.37 In mild temperatures heat pump efficiency is high but as the ambient-temperature drops the benefit of the (air-source) heat pump reduces.

6.38 The Coefficient of Performance (CoP) is a measure of the heat output for a given electrical input compared to a direct electrical heating system. The CoP is dependent upon, amongst other things, the difference in temperature of the heat source (air/ground/water) and the central heating radiators or underfloor heating (water/air) temperature required.

6.39 The heat pump CoP is a standardised measure useful to compare one product with another. The chart below illustrates the change in CoP for different refrigerants over a range of external ambient temperatures when delivering heating through a radiator (CoP is usually higher when heating through an underfloor system). [87]

Seasonal Coefficient of Performance

6.40 A Seasonal Coefficient of Performance (**SCoP**) can be more helpful as it accounts for typical seasonal temperature variations – important when using an air-source HP, but less useful for a ground-source HP.

Coefficient of Performance - Heat Pump - Radiators - Different Refrigerants

(R32, R410A, R1234yf; Coefficient of Performance vs Outside Temperature DegC, -10 to 14)

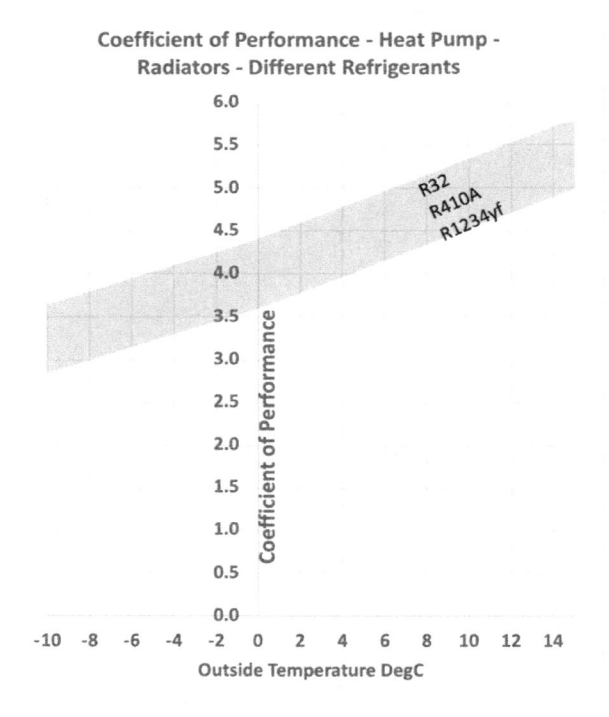

Coefficient of Performance varies with outside ambient temperature

6.41 The table below is the different SCoP for a variety of refrigerants used in a cold climate (Sweden) seasonal analysis delivering heat into radiator and underfloor heating systems UFH.

	Radiator	Underfloor Heating
R410A (Pentafluoroethane)	4.14	5.32
R32 (Difluoromethane)	4.42	5.69
R290 (Propane)	4.05	4.98
R1234yf (Tetrafluoropropane)	3.88	4.93

Plymouth's adoption of heat pumps

6.42 Plymouth City Council has saved some 402t CO2/year by the adoption of Air-Source Heat Pumps. ASHPs have been fitted into Council buildings in the Guildhall, Ballard, Elliot Terrace, and Crownhill Court. In Phase 2, ASHPs have been fitted into Pounds House, Frederick Street, Prince Rock, and Poole Farm.

The Zero Emission Boiler

6.43 The Zero Emission Boiler (ZEB) is a heat storage device capable of storing the heating requirements for a day (40kWh) and releasing it to heating or hot water throughout the day. It works in a similar way to a battery and is charged with electricity on a time-of-use principle to keep costs down and to reduce demand peaks. [88]

7. Electricity Generation, Distribution, and Storage

As well as the demand for today's generated electricity, both transport and heating/hot water, currently dominated by petroleum and natural gas respectively, will need to be done electrically.

An electric-future free of fossil fuels requires energy demand to be reduced by a third to a half.

Future electricity generated must be by zero-carbon methods.

Demand management in domestic, commercial, and industrial settings must be introduced.

Time of use restrictions and energy storage at all levels will assist in reducing carbon emissions and can relatively easily be introduced.

Interconnectors lead to a decrease in emissions, a reduction in total power market costs, less thermal generation, and less curtailment of renewable generation. [89]

Energy storage is a key enabler for combustion-free zero emission renewables.

Battery Energy Storage Systems (BESS) are relatively low cost and quick to build.

New pumped storage is expensive and takes many years to build.

Compressed air and mechanical energy storage may be attractive if they can be proven to compete with the low cost of batteries.

Grid-scale and domestic level energy storage should be incentivised.

Thermal power generation plants (gas, oil, wood/biomass) require water cooling, something less readily available due to the effects of climate change.

UK Government strategy and initiatives

7.1 The UK Government has defined its latest strategy in the British Energy Security Strategy BESS.

7.2 The Government has indicated its intention not to return to producing energy from fuels producing climate-imperilling carbon dioxide or filthy fumes from toxic pollution, instead it will take advantage of wind and sunshine as well as producing vastly more hydrogen which is easy to store and is, the government say, the 'super-fuel' of the future.

7.3 The Government will also embrace what it describes as 'safe, clean, affordable' new generation of nuclear reactors, re-establishing the UK's pre-eminence in a field where it once led the world.

PCC's CEAP initiatives

7.4 PCC will engage with Western Power Distribution (now National Grid) to establish the existing grid capacity and understand the impact of new developments.

7.5 PCC will explore smart approaches that would reduce the impact of new developments on the electricity grid.

Electricity power generation

Great Britain electricity generation 2020.

7.6 Chart shows the daily and monthly supply and demand of electrical power over the years and by source.

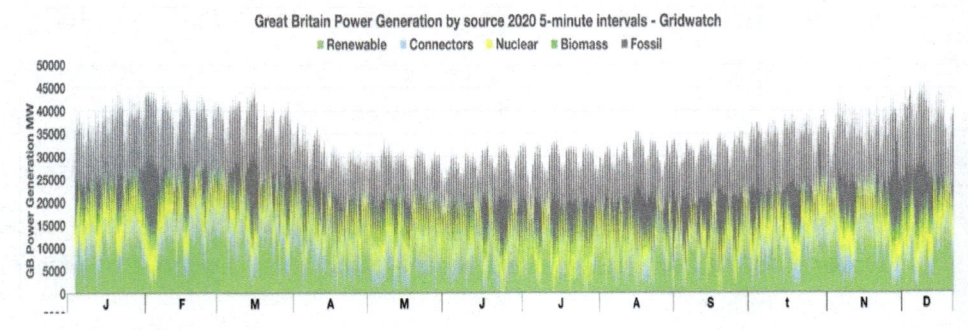

Great Britain electricity demand and source of generation for 2020 – electricity demand is variable throughout the year and variable on a daily and hourly basis. The variable supply nature of zero-carbon renewables is shown as well as the limitations of nuclear as a complementary power source

Combustion-Free Heat, Power, and Energy

Electricity transformation

7.7 Britain's electricity supply has decarbonised over the last decade. [90].

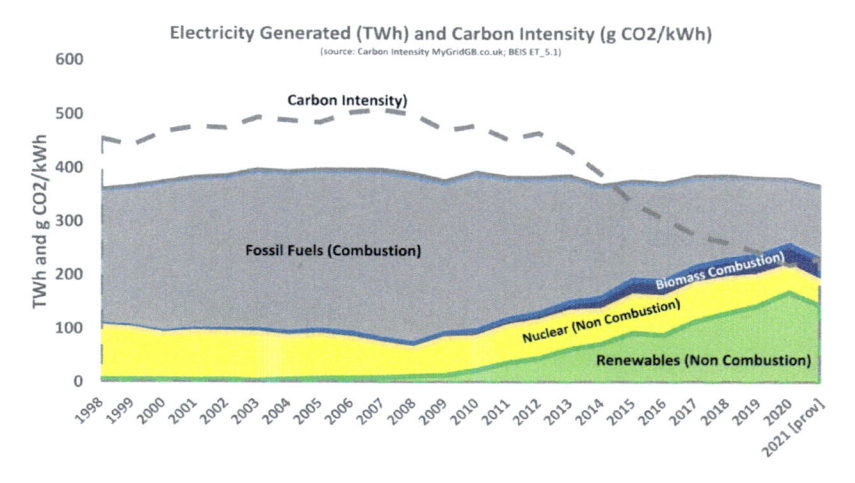

Carbon intensity of the GB national grid showing the reduction in carbon intensity driven by the significant growth in non-combustion renewables

7.8 In 2020, renewables generation reached a record 43.1% of total electrical power generation. Low Carbon generation reached 59.3%. Fossil fuel generation declined from 75.4% a decade ago to 37.7%. Coal dropped to just 1.8%, down from 28.2% a decade ago. [91]

7.9 The National Grid regularly updates its Future Energy Scenarios (FES) mapping how it could reach a net-zero carbon emissions by 2050. [92]. Four scenarios are mapped in their latest FES: Leading the Way, Consumer Transformation, and System Transformation all achieve net-zero by 2050. Steady Progression falls a long way short of the target.

7.10 By 2050, adopting the National Grid's most ambitious de-carbonisation scenario – 'Leading the Way', requires a change from today's usage to one using about 25% of combusted hydrocarbons (fossil fuels, wood, and biofuels), 65% by non-combustion renewables, 8% by nuclear and 2% imported. However, this most ambitious strategy still leaves 25% of electricity production producing 28 mtCO$_2$e greenhouse gases.

Fuel usage transformation 2020 to 2050

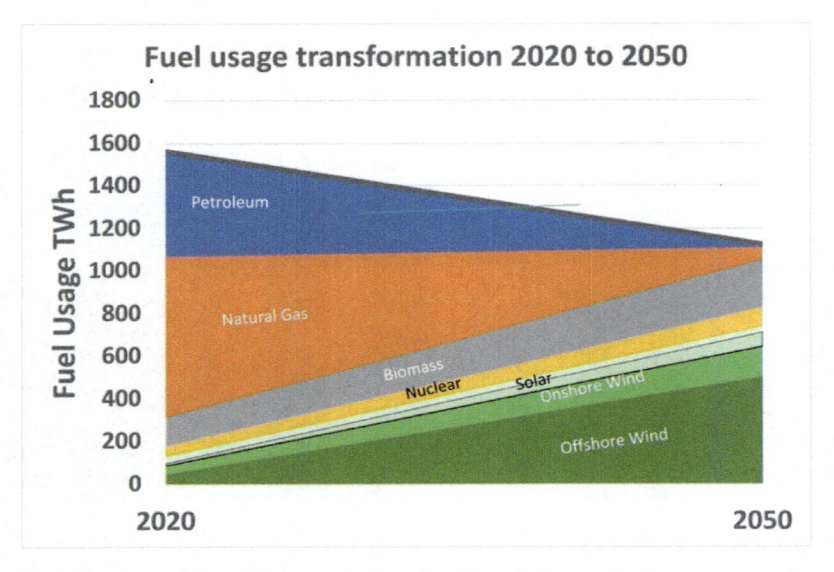

(Stacked area chart showing Fuel Usage TWh from 2020 to 2050, with layers labelled: Petroleum, Natural Gas, Biomass, Nuclear, Solar, Onshore Wind, Offshore Wind. Y-axis: Fuel Usage TWh, 0 to 1800.)

Net zero transformation requires a significant change in the energy mix.

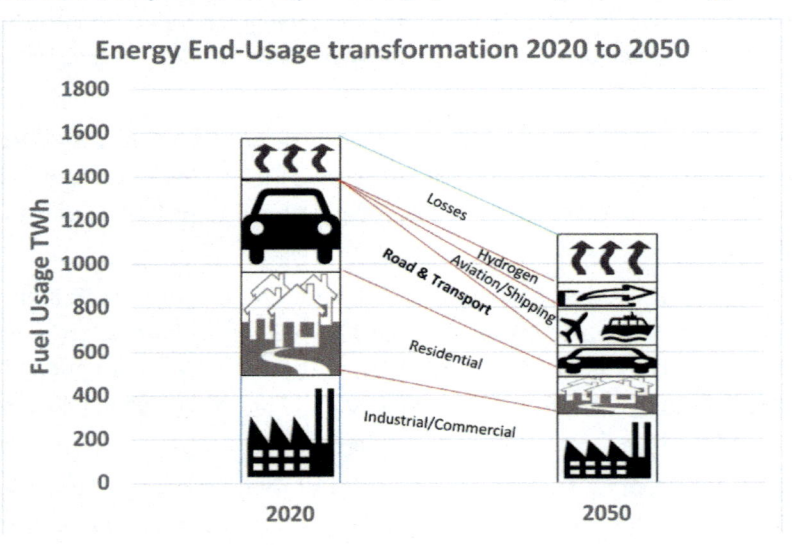

Energy End-Usage transformation 2020 to 2050

(Chart showing Fuel Usage TWh from 2020 to 2050, with categories: Losses, Hydrogen, Aviation/Shipping, Road & Transport, Residential, Industrial/Commercial.)

Net zero transformation requires a reduction in energy use from every user.

7.11   2050 requires a two thirds reduction in energy used and a seven-fold increase in wind power generation to reach net zero. Total annual energy demand if provided by all electric wherever possible ~ 1,577 TWh in 2020 or 1134 TWh in 2050.

7.12   The National Grid, responsible for the UK's gas and electricity transmissions, will be partly returned to public ownership by 2024. This is to aid the development of the grid as energy moves from gas to all-electric. [93]

Combustion-Free Heat, Power, and Energy

Peak demand and variability

7.13 **2020 Electricity power demand** – between 20 and 45GW depending on time of day, day of the week, and month of year as well as ambient temperature and solar radiation. [94]

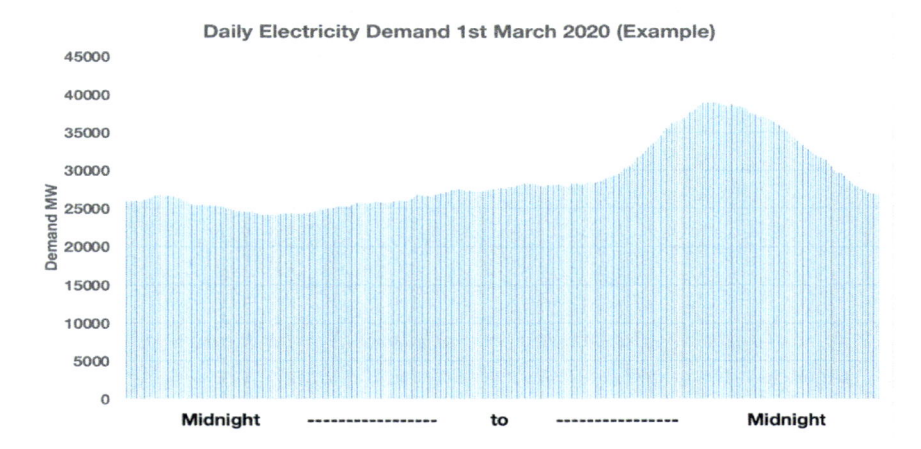

Daily electrical demand for 1st March 2020 as an example of daily variability. Total electricity used in the day was 697,181MWh.

7.14 Typical **natural gas usage** on an annual and typical daily basis

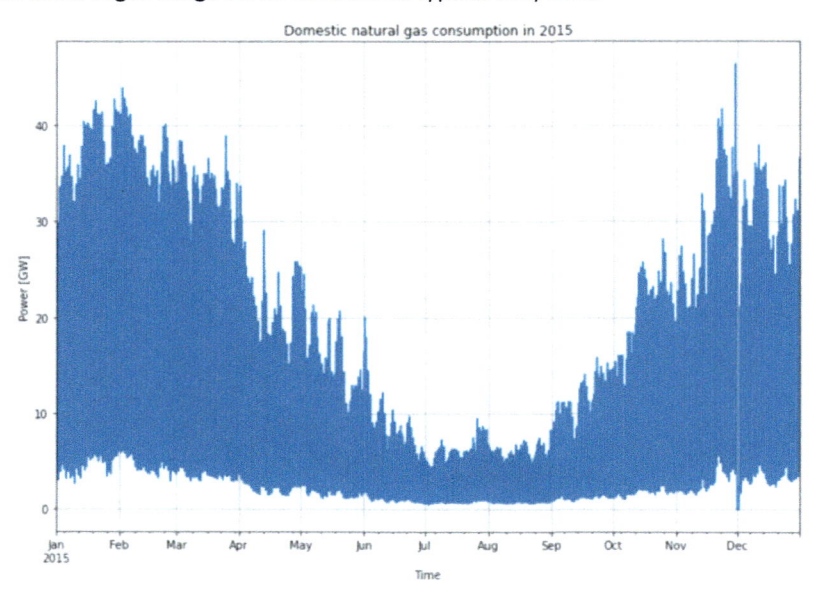

Autumn, winter, and spring have significant heating demands.

- 50 -

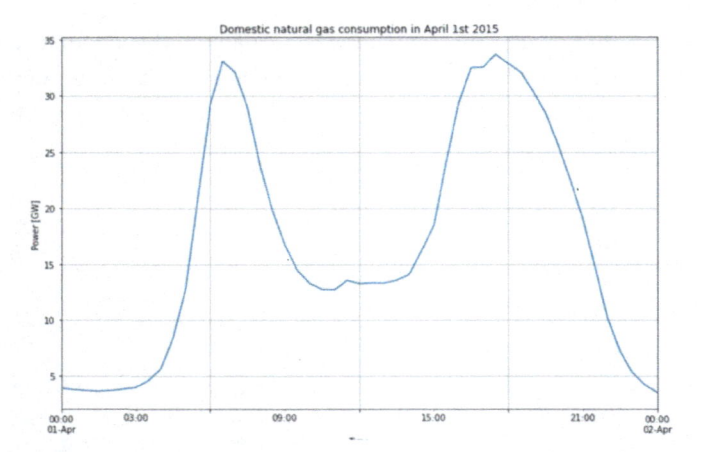

Climate Action Plymouth – Sustainable Power and Energy Group

Annual and daily gas demand for a day in April as an example of daily variability. Because natural gas is more easily stored, the importance of hourly need is not so critical, but it will be important when heating transfers from gas to electric.

7.15 Peak electricity usage can significantly increase both costs and carbon emissions as peaking plants tend to be less efficient, high-emission, gas or diesel fuelled. Mitigating actions include battery storage, shifting peak demand away from evening cooking, and using time-of-use pricing, particularly useful for electric vehicle charging and for 'wet' usage such as dishwashers and washing machines. [95]

Renewables capacity

7.16 Installed capacity of renewable generation assets by technology 2000-2019

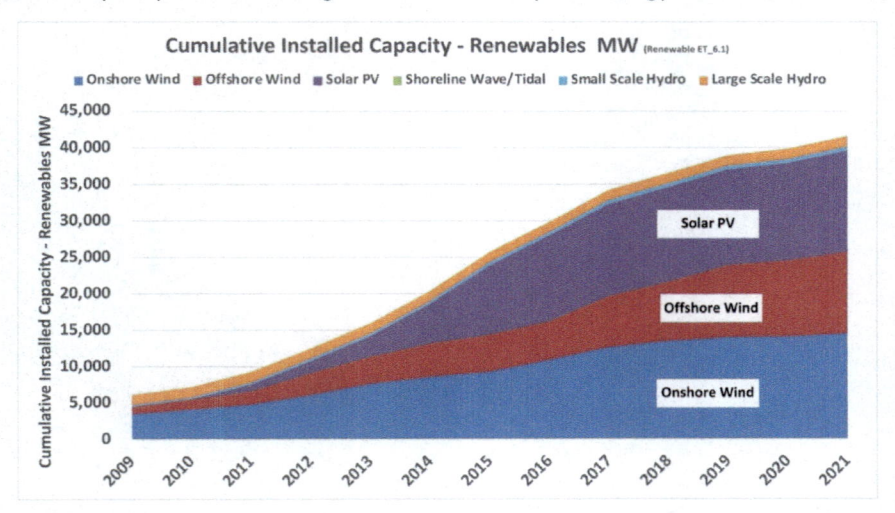

Installed capacity of non-combustion renewables showing significant growth over the past decade.

- 51 -

Electricity National Grid Carbon Intensity

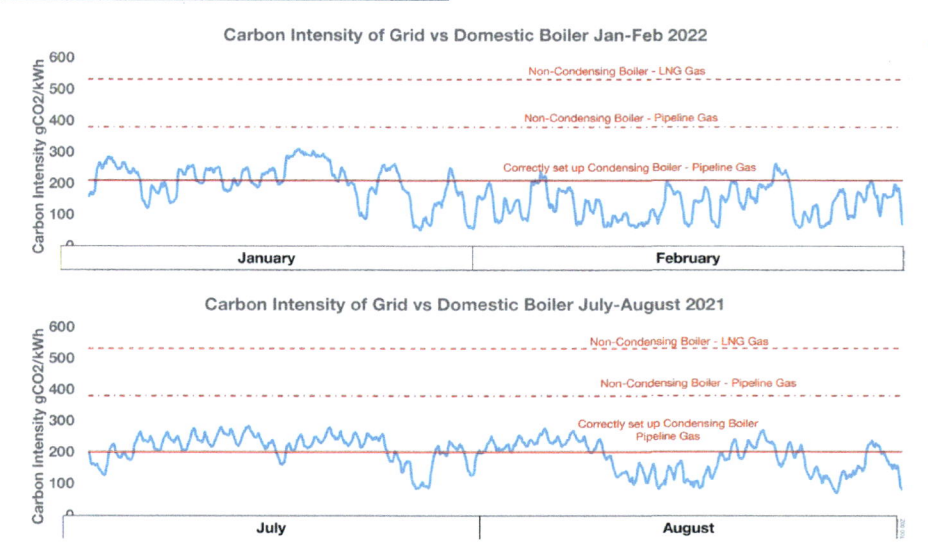

Carbon intensity of the national electrical grid during two different months, one in January-February, one in July-August. These are shown against the carbon intensity of natural gas boilers using pipeline gas and using liquified natural gas LNG.

7.17 The 2020 electricity grid will deliver lower carbon intensity than a gas boiler, and as the grid gets even greener, this will become even lower.

7.18 Carbon Intensity of the electricity national grid is measured by the National Grid ESO, in partnership with Environmental Defense Fund Europe, University of Oxford Department of Computer Science and WWF. They have developed the world's first carbon Intensity forecast with regional breakdown. [96]

7.19 Domestic boiler carbon intensity depends upon the age and type of the boiler and the source of the gas – pipeline or LNG. LNG now makes up 26% of UK gas supply [DUKES 97].

7.20 The correctly set up condensing boiler, delivering heating water at ~60°C to maximise condensing efficiency, and running from pipeline gas, has a carbon intensity between 210 and 230 gCO_2/kWh, A non-condensing older boiler may have a carbon intensity of as much as 380 gCO_2/kWh. [98] LNG has a carbon intensity typically 40% worse that pipeline gas due to the power required to liquify the gas, transport it, and re-gas it. This leads to a carbon intensity of natural gas when used in a domestic boiler of over 500 gCO_2/kWh.

7.21 There are now more condensing boilers than non-condensing boilers, but many are retrofitted into older radiator set ups requiring a higher water temperature, up to 80°C. This prevents condensing and negates the efficiency gain.

Regional energy grids

7.22 The electricity distribution network is managed by six companies:
- Electricity North-West Ltd (ENWL),
- UK Power Networks (UKPN),
- Northern Powergrid (NPg),
- SP Energy Networks (SPEN),
- Scottish and Southern Electricity Networks (SSEN),
- Western Power Distribution (WPD) now National Grid.

7.23 Ofgem controls the pricing of these Distribution Network Operators (DNO) and has revealed a five-year vision to build sustainable and affordable grids supporting low carbon electricity. [99]

Levelised Cost of Electricity (LCOE)

7.24 The Levelised Cost of Electricity (LCOE) is the discounted lifetime cost of building and operating a generation asset, expressed as a cost per unit of electricity generated (£/MWh). It covers all relevant costs faced by the generator, including pre-development, capital, operating, fuel, and financing costs. This is sometimes called a life-cycle cost, which emphasises the "cradle to grave" aspect of the definition." [100]

7.25 Assumptions on fossil fuel costs are a 2019 baseline. (*Wholesale gas prices soared by 200% in 2021. Further rises occurred in 2022*)

7.26 Nuclear unchanged from 2016 BEIS report.

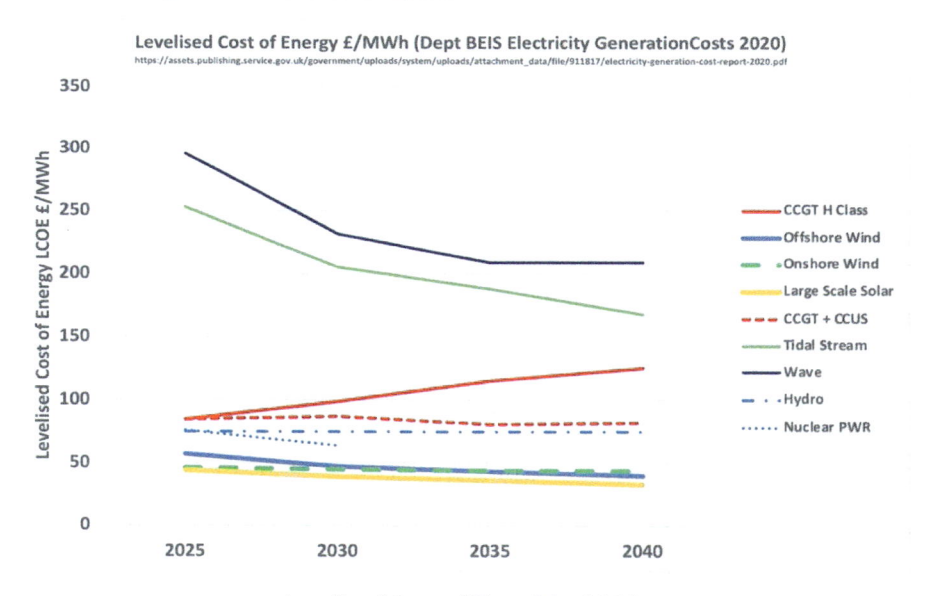

Levelised Cost of Electricity LCOE

Electricity interconnectors

7.27 Electricity Interconnectors are cables transmitting electricity between the UK, Ireland, France, the Netherlands. Interconnectors to Belgium and to Norway are under construction. Approximately 12GW are planned or under construction by 2025. [101]

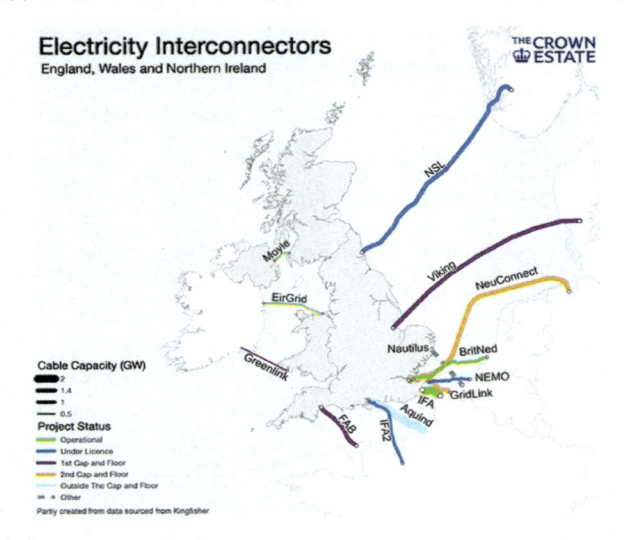

Illustration showing the electricity interconnectors operating today plus the ones under construction or under development

X-Link Interconnector Morocco-UK [102]

7.28 The X-Link will be the longest interconnector in the world. At 3,800km it is a very long high-voltage DC cable running from a solar and wind farm (complete with battery back-up) in Morocco to North Devon, where the electricity produced, 3.6GW, will be fed into the Grid. The solar and wind farms in Morocco have yet to be built - power is planned to be delivered by the end of this decade. China has installed quite a few of these schemes although none quite as long as this one.

7.29 Cables will run through France, Spain, and Portugal's Exclusive Economic Zones (EEZs). Energy losses are low (~13%) but capital costs are quite high (~£18bn). Power produced and costs are about the same as Hinkley C nuclear power station.

UK-Germany Neuconnect Interconnect Project

7.30 Funded by the European Investment Bank, Neuconnect is a €2.8bn high-voltage DC link between UK and Germany. The submarine cable will be 725km with a capacity of 1.4GW operating at a DC voltage of 525V.

7.31 Further Interconnectors joining British and Dutch wind farms are proposed, helping to tackle intermittency. This will provide a further 2GW of capacity.

Interconnectors proposed UK-Morocco and UK-Germany

7.32 Initial funding is provided by the Government's Innovate UK organisation; longer term funding is not clear, but X-links says "it does not expect to require subsidy or financing from the UK government. Using the contracts for difference (CfD) mechanism, it will provide energy at £48/MWh CfD. This is below the Department for Business, Energy & Industrial Strategy (BEIS) central forecast, which will provide savings for consumers."

7.33 Morocco has a gas dispute with Algeria and so has increased its reliance on coal. X-Link proposal to spend £18bn of investors' money on an overseas electricity generation scheme that UK says it will use for the balancing mechanism rather than for dispatchable energy. This seems a hard sell.

Environmental impact of interconnectors

7.34 Interconnectors are invaluable in enabling the balancing of generation and demand Doing so will help avoid using higher carbon peaking plant.

7.35 The environmental impact of the interconnectors varies with the source of the electricity generated.

Energy Storage

7.36   Energy storage can take many forms and deliver different benefits. It can be at the grid level, local level, or at the domestic level.

7.37   The difference between energy and power is important when considering storage. Energy is measured in kWh and is a measure of how much power for how long - it is a measure of the capacity of the energy storage system. How quickly the energy can be delivered is power, measured in kW.

Energy storage usage

7.38   Energy storage is used for
- **Power arbitrage/load levelling** – storing energy when supply exceeds demand (or when costs are low) and supplying it when demand is high (or when prices are high). Power arbitrage can be extremely useful in micro grids powered by renewables when the variation in supply can be significant – e.g. solar power and intermittent wind. At the domestic level, energy storage can be used to avoid peak times as well as shifting electricity usage to low carbon hours.
- **Firm capacity/peaking capacity** – energy storage supplies power during the highest demand periods, often for quite short timescales.,
- **Operating reserves** - primary frequency regulation, spinning reserve, ramping/load following, - these ensure a reliable grid delivering a stable frequency power supply
- **Black start** – necessary to restart generators when grid power is unavailable.

7.39   Energy storage is not currently used for long term capacity. An average day of electricity in the UK is about 750GWh. For a future all-electric scenario in 2050, around 3,000GWh per day will be used once the current gas usage is replaced by electric. However, energy storage will grow to enable longer term storage for balancing daily variations in supply and demand, managing several days of oversupply/undersupply, and, with some technologies, managing seasonal differences.

7.40   The UK Government is funding projects to deliver **'Longer Duration Energy Storage'**. [103 104] Contracts have been placed for heat/thermal batteries, ultra-high temperature energy storage, phase-change thermal storage, pumped thermal energy, lithium-sulphur flow batteries, zinc-metal and copper-zinc batteries, marine pumped hydro, thermal and compressed air storage technologies, and green hydrogen storage. Of particular interest to Plymouth is the proposal by RheEnergise to develop "High-Density Hydro®, a cost optimised energy storage solution to create stable, dispatchable power grids from ultra-low-cost renewable energy". RheEnergise will deploy a 250kW/1MWh demonstrator at a mine, close to Plymouth by mid-2024. [105]

Energy storage technologies

7.41 Energy Storage technologies include pumped storage (pumping water up to a reservoir and releasing it through a turbine to generate electricity). Pumped storage was introduced because both coal and nuclear were slow to respond to demand fluctuations. Pumped storage is well suited to renewables generating fluctuations.

7.41.1 Perhaps the best-known pumped storage is Dinorwig power station (which opened in 1984 after ten years in construction) in Wales. It is capable of up to 1.73GW of power with a storage capacity of 9.1GWh. It was built in the days of the Central Electricity Generating Board CEGB.

7.41.2 Three other pumped storage facilities exist. Two in Scotland – Cruachan (440MW) and Foyers (300MW), and a second in Wales – Ffestiniog (360MW). Other pumped storage facilities were planned but not built.

7.41.3 Pumped storage is neither cheap (>£500m) nor quick (5-8 years) to bring on-line. [106]. The Government does not include pumped storage in the 'Contracts for Difference' funding mechanism.

7.41.4 In total (July 2022), the UK has 25.8GWh of pumped storage and 1.6GWh of battery storage. [107]

Dinorwig Pumped Storage Power Station

7.41.5 The UK's largest **battery energy storage system (BESS)** is based in Essex. It deploys a 320MW/640MWh system delivering fast reacting power. It is expandable to 1.3GWh (about a seventh of Dinorwig but much quicker to build). [108]

7.41.6 The largest BESS currently is in California and is 400MW/1.6GWh. Other BESS being planned include a facility up to 5GWh. [109]

7.41.7 The National Grid, in its Future Energy Scenario (FES), envisages 65GWh in 2030 and up to 194GWh by 2050 (132GWh non-battery and 65GWh battery) [110]. The UK

Government relaxed planning rules to maximise the UK's renewable energy storage in July 2020. [111]

Artist's impression of the UK Gateway largest Lithium-Ion Battery Energy Storage System BESS

7.41.8 **Domestic level battery energy storage**, the Government says, 'is not only an attractive clean option to grid supplied electrical energy but is on the verge of offering economic advantages to consumers, through maximising the use of renewable generation or by 3rd parties using the battery to provide grid services.' [112]

7.41.9 In 2022 domestic battery energy storage is approximately £500 per kWh. The packs are quite compact and can be fitted inside or outside the home. Modules of between 6 and 14kWh seem readily available; there seems little reason why a day's worth of energy should not be stored by a significant number of homes. Assuming 28kWh was available to 50% of households (~14m) [113] then nationally a store of approximately 400GWh is possible.

Domestic-level commodity Lithium-Ion Battery Energy Storage System BESS. Available from a few kWh to 135kWh.

(Picture shows a 13.5kWh battery storage Photo Brian Piper)

Combustion-Free Heat, Power, and Energy

7.41.10 By 2030, all new cars on the road will be plug in. Many models are now available with **Vehicle-to-Grid V2G** capability allowing the car to feed electricity back to the grid. The National Grid supports this strategy and OFGEM have conducted trials. [114 115] V2G, when mature could enable over 120GWh of energy storage.

7.41.11 **Flow batteries** are electrochemical devices of which the Redux system is the most common. Redux flow batteries use a liquid electrolyte which is cycled through the battery. A liquid phase reduction-oxidation reaction produces the power. The electrolyte is then pumped back through the electrodes to the tanks. The flow battery can be cycled indefinitely. It is quite low energy density compared to a lithium-ion battery. The largest flow battery in the UK is in the Energy Superhub Oxford delivering 5MWh as a front end to a 50MWh lithium-Ion system. [116]

7.41.12 **Heat batteries** store energy in a 'phase change material' rather than in a hot water tank. They are non-toxic and compact. They provide a credible option to a hot water tank when used to store heat derived from solar, wind or off-peak electricity. Extracting heat involves passing cold water through the heat batteries heat exchanger.

7.41.13 **Thermal energy storage** utilises the heat that can be stored within inexpensive silica sand. Storing heat at up to 1200°C for many days, the energy is released when the sand passes through a heat exchanger used to pressurise a gas that then drives turbo-generating machinery. A prototype is being tested at the National Renewable Energy Laboratory; it has the potential to be a low cost, high-efficiency cost-effective system. [117]

7.41.14 **Compressed air** storage as a concept is more than 40 years old and was considered as a solution to complement nuclear power. It has been used for two commercial scale plants as an alternative to pumped hydro storage. [118 119] As with compressed air transport vehicles, it is likely that this technology has been overtaken by Li-Ion Battery Energy Storage Systems.

7.41.15 **Compressed CO₂ storage** has a higher density than compressed air. CO_2 is liquified under pressure reducing volume significantly. [120]

7.41.16 **Mechanical energy storage** includes weights and pulleys, flywheels, hydraulic pistons, buoyancy energy stores, etc. [121]. None of these technologies seem destined to compete successfully against BESS.

7.41.17 **Hydrogen,** when produced by excess renewable electricity for storage and then used in zero-emission fuel cells may well provide a successful energy storage system. See the chapter on Hydrogen for more details.

Environmental impact of energy storage [122 123 124 125]

7.42 The environmental impact of pumped storage is like the other projects that require dams. There are impacts on wildlife habitats and on migration paths. Damming rivers has a considerable negative impact on the rivers and their wildlife. Casualties amongst the fish population can also be expected. Construction impacts on the livelihoods of those displaced can also be detrimental.

7.43 Utility-scale battery storage can significantly reduce the emissions of GHG and toxic pollution from fossil fuel use and is a significant enabler of intermittent renewables. However, batteries use a wide range of minerals involving mining, manufacturing, transportation, and disposal.

7.44 Utility scale batteries, rather than the Nickel Manganese Cobalt (NMC) vehicle batteries, generally use Lithium Ferrous Phosphate (LFP) that have electrodes of non-toxic iron and graphite.

7.45 Aluminium is also another common metal used in li-ion batteries. Aluminium production can involve CFC-11 that leads to ozone depletion.

7.46 Lithium itself is not toxic although its extraction from lithium brine can cause depletion of water reserves. Lithium is also produced from open pit hard rock mining that can lead to toxic soils and dust.

7.47 Phosphate salts are less likely to leach out during disposal than metal oxides.

8. Fossil fuels and biomass

In the UK, the fossil fuel used in electricity generation or in space heating and hot water is predominantly natural gas. It is carbon intensive and emits toxic NOx and particulates.

Natural gas (predominantly methane) for space heating/hot water should be phased out and replaced by electricity, generated by renewables and not by natural gas.

Liquified Natural Gas (LNG) is substantially more carbon intensive than pipeline gas and should incur a higher tax to discourage its use. Exports of North Sea pipeline gas should be restricted. No new gas exploration licenses should be granted. Natural gas storage should be restored/expanded enabling pipeline gas to be used in place of LNG.

A domestic gas boiler burning pipeline gas emits a similar amount of CO_2 per year as around seven transatlantic flights (or around ten flights if burning LNG).

Wood and biomass are used both for electricity generation and for space heating. Wood burning and biomass emit greenhouse gases and should be discouraged. Burning wood and biomass emits particulates, NOx and other harmful gases. Biomass is a significant contributor to the UK's claim to renewable energy, but it should no longer be considered a 'renewable'. Biomass does not contribute to the reduction of greenhouse gases and is not likely to achieve carbon neutrality before the end of the century, if at all.

Oil is used for space heating and hot water in rural communities where there is no access to the gas grid. Coal has almost entirely been phased out.

Carbon Capture Utilisation and Storage (CCUS) aims to remove carbon from the emissions of power plants and industrial processes. Only one CCUS plant remains in operation in 2022. There remain significant technical and commercial challenges to CCUS. It seems highly unlikely that CCUS can contribute much, if anything, to CO_2 reduction by 2030.

UK Government strategy and initiatives

8.1 The UK Government in its British Energy Security Strategy (BESS) estimates that natural gas will still make up 25% of our energy usage in 2050. There are still 7.9bn barrels of oil reserves and 560bn cu.m of gas remaining in the North Sea. Another licensing round for North Sea oil and gas is expected in autumn 2022; fracking is currently (Oct 2022) banned again until proven to be safe. The government has an ambition for carbon capture of 10mt of CO_2 by 2030

Natural gas

8.2 The late 18th century saw town gas introduced for lighting in a house. Within 30 years, towns would be lit by gas. Gas was later used for cooking. Post First World War, gas began to be used for home heating. North Sea or natural gas came ashore by pipeline in 1967. A ten-year conversion programme followed to convert domestic appliances from town gas to natural gas.

8.3 'Wet' natural gas is predominantly methane (CH_4) at 60-90% with up to 20% made up from ethane, propane, and butane. Other constituents include carbon dioxide, oxygen, nitrogen, hydrogen sulphide, and rare gases. [126].

8.4 Processing removes ethane, propane, and butane, as well as gases such as carbon dioxide and nitrogen which would otherwise reduce the calorific value of the natural gas. Sulphur is also removed and replaced by an odorant to assist in leakage detection. This leaves 'dry' natural gas consisting almost entirely of methane (85-90%) with the balance being ethane and nitrogen.

8.5 Natural gas is used by most homes in the UK and Netherlands. Canada and USA use natural gas for about half of their homes. France and Germany for about 40%. Scandinavian countries do not use natural gas for heating or hot water. This disparity is linked to the availability and price of gas to the consumer. For the UK and Netherlands, proximity to the North Sea gas fields is significant. Norway has an abundant supply of hydro generated electricity and so can export its natural gas.

Combustion-Free Heat, Power, and Energy

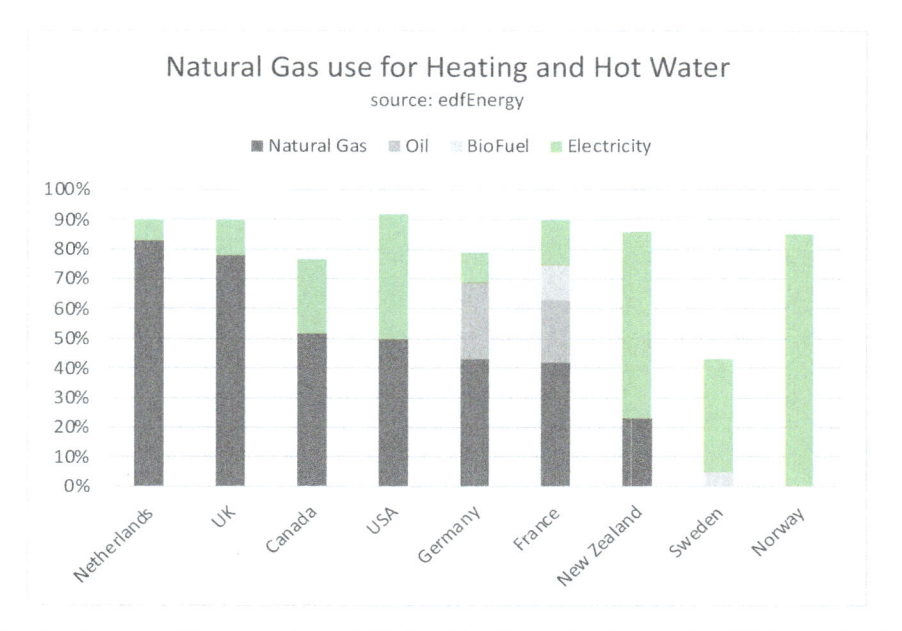

Natural Gas use for Heating and Hot Water
source: edfEnergy

Heating homes with natural gas is high in the Netherlands and the UK, less so in the rest of Europe, and not used in Sweden or Norway.

Source of gas

8.6 Natural gas is transported either by pipeline or, if liquified, by ship. Pipeline gas made up 74% of natural gas supplies used by the UK in 2020; LNG accounted for the other 26%. Imports made up 63% of natural gas used in the UK in 2020; North Sea accounted for 37%.

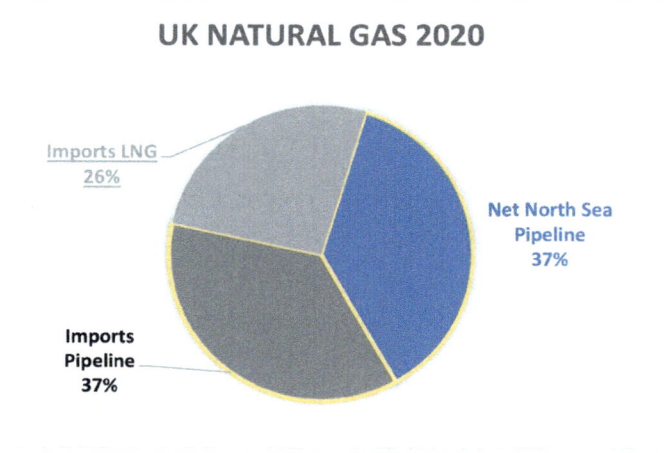

UK NATURAL GAS 2020

- Imports LNG 26%
- Net North Sea Pipeline 37%
- Imports Pipeline 37%

LNG is not only more expensive than pipeline gas but also has much higher carbon emissions. It now makes up a significant proportion of the gas used in the UK.

Pipeline Gas (2020)

8.7 North Sea/continental shelf and onshore natural gas production is 438,330 GWh, of which 51,107 GWh (12%) is used for its own production (drilling, production, and pumping). 105,903 GWh (24%) of North Sea pipeline gas is exported.

8.8 Total UK natural gas for the UK from the North Sea by pipeline is 281,320 GWh.

8.9 Pipeline Imports to the UK are from Norway (263,495 GWh), Netherlands (11,073 GWh), and Belgium (3,554 GWh) – Total 278,122 GWh.

8.10 Total net pipeline gas is therefore North Sea plus pipeline imports 559,442 GWh (2020 figures).

Pipeline Gas	GWh
North Sea production	438,330
Drilling, production, pumping	(51,107)
Net production	387,223
North Sea exports	(105,903)
Net from North Sea to UK	281,320
Pipeline imports	278,122
Total pipeline gas to UK (2020)	559,442

8.11 The **Carbon intensity** of pipeline natural gas accounts for both CO_2 and 'lost' methane.

8.12 About 13% of gas losses occur in production, transport (compressor units), and gas leaks from transmission, storage, and distribution as well as in power/heat generation. [127]

8.13 When considering the GWP of methane, the CO_2e from lost methane is equivalent to the CO_2 produced by production, processing, transmission, and storage.

Liquid Natural Gas LNG

8.14 Liquified Natural Gas LNG has been used in Great Britain since the sixties but was mostly used for 'peak shaving' rather than for general use.

8.15 Liquification is a carbon intensive task. [128] Raw feed gas must be scrubbed of hydrocarbon liquids and dirt and then treated to remove any hydrogen sulphide and carbon dioxide. The gas is then cooled to condense out any water and then dehydrated to remove any remaining water vapour. If mercury is present, this is also removed. The gas is then filtered leaving predominantly methane with just small amounts of light hydrocarbons, mostly ethane.

8.16 Liquefaction of LNG is via either single or multi-stage refrigeration after which the liquid gas is stored at -160°C in an insulated container as a 'boiling' liquid.

Combustion-Free Heat, Power, and Energy

8.17   Transport of LNG is by dedicated LNG ships, either operating as a 'train' or more recently selling on the spot market. Propulsion of LNG tankers is by steam turbine heated by a mix of 'boil-off' gas and heavy fuel oil, or via dual-fuel reciprocating diesel/gas engines. At the destination the gas is heated to speed up boil off.

8.18   LNG in 2020 delivered 26% of the natural gas into the UK grid. A total 200,066 GWh of LNG by ship arrived from Qatar (48%), USA (27%), Russia (12%), Trinidad and Tobago (6%), and others (7%) – 2020 figures [129].

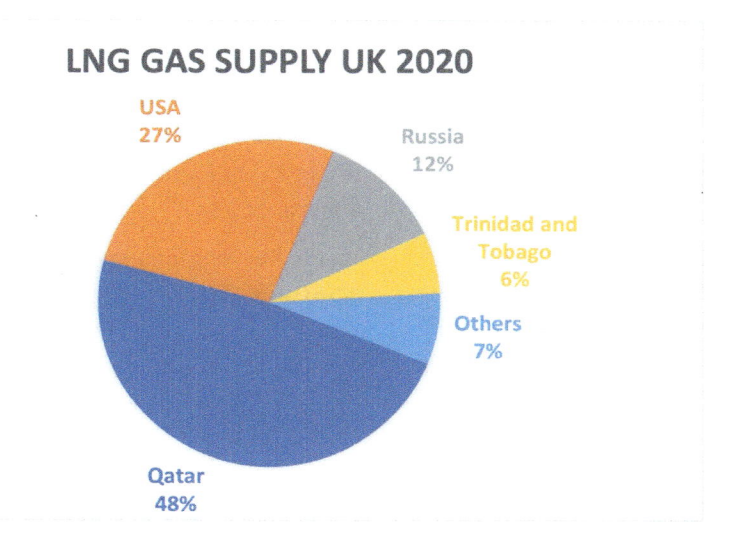

LNG GAS SUPPLY UK 2020

- USA 27%
- Russia 12%
- Trinidad and Tobago 6%
- Others 7%
- Qatar 48%

8.19   LNG has significantly higher carbon intensity (GHG emissions) than pipeline gas, sometimes comparable in intensity to other oil projects. [130]. Because CO_2 is stripped out of LNG prior to liquefaction, combustion of LNG is lower in GHGs than pipeline gas. However, overall GHG emissions are higher for LNG due to the energy used for liquefaction as well as the need for CO_2 stripping and venting before liquefaction.

8.20   Natural gas is increasingly a spot market commodity and as such could come from any of a hundred or more sources, including from shale gas, via pipeline or LNG, from Russia or the Far East, etc. Studies have shown that greenhouse gases emissions from 'well to city/power station', can vary by up to seven times, making it quite difficult to establish at any one time what the emissions are for gas generated electricity or heating. [131]. For this report, the carbon intensity of LNG is taken as 40% greater than pipeline gas. [132]

Environmental Impact of natural gas [133 134]

8.21 Natural gas is a fossil fuel consisting primarily of methane. When burnt it produces carbon dioxide (CO_2) and unburnt methane (CH_4).

8.22 Natural Gas is a fossil fuel and is not renewable.

8.23 When combusted, natural gas produces toxic emissions – Nitrous Oxides (NOx), Carbon Monoxide (CO), and Sulphur Oxides (SOx), plus a low level of particulates - causing air pollution, particularly in towns and cities.

8.24 Great Britain's domestic gas boilers produce around twice as much carbon emissions as gas used in power generation.

8.25 Gas production sites suffer erosion, contamination, and wildlife and biodiversity issues.

8.26 Hydraulic fracturing (fracking) can cause earth tremors and contamination of water supply with methane and/or hydraulic fluids.

8.27 In the UK there have been typically 30 unintentional deaths per annum due to carbon monoxide (CO) poisoning. [135] There are on average nine (9) natural gas explosions in UK dwellings per year.

Wood burning and biomass

Biomass

8.28 "Biomass is a renewable energy source, generated from burning wood, plants and other organic matter, such as manure or household waste." [136]

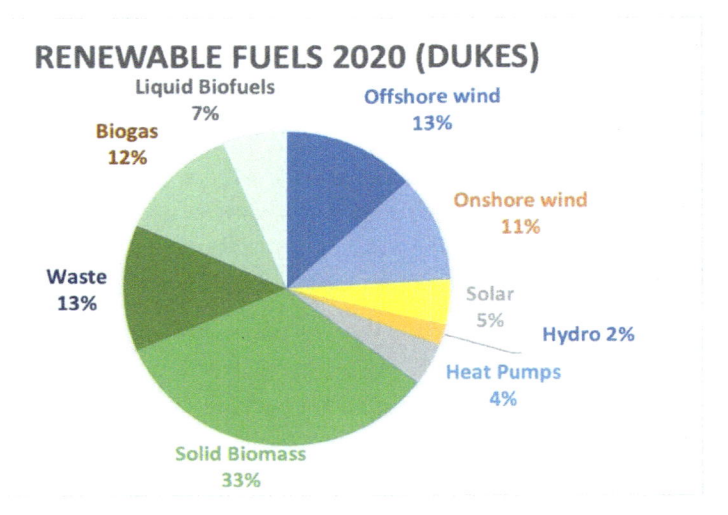

RENEWABLE FUELS 2020 (DUKES)

- Liquid Biofuels 7%
- Offshore wind 13%
- Biogas 12%
- Onshore wind 11%
- Waste 13%
- Solar 5%
- Hydro 2%
- Heat Pumps 4%
- Solid Biomass 33%

8.29 Solid biomass, including wood, waste wood, animal, and plant biomass, represented 33% of total renewable demand in 2021. Two-thirds went to electricity generation, the remainder to heat. [137]

8.30 The Government's *Biomass Policy Statement November 2021* considers that "Biomass has a role to play in achieving net zero, in delivering greenhouse gas removals (GGRs) at scale and changing the way we use our land to support carbon sequestration and clean energy production among other things". "Biomass is a vital resource for the key green technologies and energy carriers necessary for net zero: low carbon electricity, hydrogen, carbon capture, and bioenergy." [138]

Environmental impact of biomass

8.31 Biomass burning releases carbon dioxide and so is not a 'zero-carbon' energy source. Biomass burning is considered 'net-zero carbon' provided the carbon dioxide emitted is the same amount as was absorbed over the months and years that the plant was growing. Biomass is considered 'sustainable' if new plants continue to grow in place of those used for fuel. [139]

8.32 Biomass sustainability arguments ignore both the need to replenish the soil with organic material, which does not happen with biomass removal, and the length of time it takes to grow back the trees and plants necessary to recover the carbon. With a tree growth time of 30-100 years, biomass is carbon positive for the near future, but it is now, in the early part of the 21st century, that we must do the most to reduce carbon emissions. [140]

8.33 Green grass mills, growing grass to produce methane to use in combustion within domestic boilers, still produce toxic emissions. [141]

Wood pellets

8.34 Wood pellets are used for biomass burning to increase the energy density of wood as a fuel. Wood pellets produce almost 4x the carbon per energy produced than natural gas and 1.5x that of coal. [142]

8.35 Wood pellets are produced in the USA. There is no USA domestic market for wood pellets, and they are produced specifically for the UK/EU. There are some sawmill residues (~20%) and some thinning (~14%) but most of the timber is grown and felled specifically for wood pellets for biomass. [143]

8.36 Wood pellet fuelled electricity generation is increasing in today's energy mix – old coal fired power stations are converted to wood pellets. Tilbury B, Ironbridge, Drax, Eggborough, and Alcan Lynemouth have all converted from coal to wood. These power stations will use 31m tonnes of wood pellets made from 62.9m tonnes of wood – about 6x the total UK wood production and about 4x the total global wood pellet production in 2010. [144]

Environmental impact of wood pellets

8.37 Promoting wood as an energy source for electricity generation will, in any realistic scenario, increase carbon emissions for decades to come regardless of the forests they are grown in and how sustainably they are managed – every kWh of wood burnt doubles the amount of carbon emissions over the next thirty years compared to an equivalent energy from fossil fuels! [145]

Domestic wood burning stoves

8.38 Domestic wood burning stoves have become a must-have for many middle-class homes, restaurants, and hotels but are rarely, if ever, a necessity.

8.39 Only wood pellet back-burning stoves are considered for the Renewable Heat Incentive RHI. [146]

8.40 Wood burning stoves produce high levels of NOx and particulates from the chimneys and from opening the door to refuel. They should be banned from urban areas. [147 148]

Carbon Capture Utilisation and Storage (CCUS)

8.41 Carbon Capture Utilisation and Storage (CCUS) removes carbon dioxide from the emissions of power generation and industrial process.

8.42 Once captured, CO_2 may be [149]:

- used for 'Enhanced Oil Recovery' (EOR) enabling greater fossil fuel extraction,
- used as a feedstock to produce fuels, chemicals, and products such as concrete,
- used in food and beverages,
- used as a refrigerant for use in heat pumps and refrigerators,
- used in fire protection as an extinguishant.
- transported (pipeline, liquefaction, etc.) to a secure deep geological **storage**,

8.43 Storage in a deep geological facility assumes that the gas can be transported there and stored without leakage, forever. The most likely transport method is by energy intensive liquefaction.

8.44 Reusing the CO_2 does not remove it from the atmosphere: it simply delays its emission.

8.45 There is little momentum behind CCUS despite parts of the USA providing financial incentives for equipping power plants.

8.46 Environmental concerns focus on pressure build-up from the injection of CO_2 into the underground storage rocks and on any CO_2 that seeps out. [150]

8.47 Given that the CO_2 should remain in the underground storage for thousands, probably tens of thousands of years or more, long term stewardship is a concern.

8.48 Handling of CO_2 prior to injection into underground storage is likely to be at or near its critical pressure and it behaves similarly to a liquid. The UK Health and Safety Executive (HSE) considers that handling CO_2 in such circumstances, although not requiring COMAH (Control of Major Accidents Hazards) regulations, should be undertaken with great care. [151]

8.49 Due to the specific geology of Iceland, CCUS has an operational facility at the Orca Plant. The installation has demonstrated that with the right geology and access to large amounts of water, CO_2 injected into subsurface basaltic rocks can be mineralised within a period as short as two years.

8.50 Whilst showing promise, CCUS remains quite some way from the scale necessary to impact CO_2 emissions and quite some way from showing economic viability.

9. Renewables (zero carbon in operation)

Non-combustion zero carbon renewables have increased significantly over the past decade, but their growth must be further accelerated rapidly to ensure carbon emissions stop rising from 2025 and to ensure that net-zero is reached by 2050.

Energy storage should be a condition of license to encourage higher capacity factors.

Onshore wind is the quickest renewable energy to bring to market. 30GW of onshore wind by 2030 is achievable. Offshore wind should reach 50GW by 2030. Bladeless wind should be encouraged near populated areas.

Solar farms, backed by energy storage, offers a quick and least expensive renewables growth option. Done sympathetically, solar farms can be beneficial to biodiversity whilst making use of poor-quality agricultural land.

Hydro power generation has limited growth potential and has a poorer rate of return on investment than solar or wind.

A Severn Estuary tidal study has been relaunched by the Government. Any scheme will be unlikely to deliver anything before 2030 at best, be expensive, and not be popular with environmental groups.

Tidal stream is in its infancy, has limited potential for growth, and has significant development issues to overcome.

Wave energy offers significant potential for energy harvesting but has failed to achieve any commercial success. Trials continue but seem unlikely to be at a scale to make a useful contribution before 2030, at best.

England has limited potential for economic geothermal power production and limited potential for hot water with today's drilling capabilities. With deeper drilling (>7km), larger areas of the country could offer the possibility of district level geothermal hot water. Funding for further development should be provided.

UK Government strategy and initiatives

9.1 The UK Government in its **British Energy Security Strategy (BESS)** projects up to 50GW of **offshore wind**, including 5GW of floating offshore, by 2030. This is an increase over the **Green Industrial Revolution (GIR)** of 40GW of offshore wind, including 1GW floating, by 2030. 70GW of solar capacity to be available by 2030. VAT will be removed from solar panels. £20m per year will be ringfenced for tidal stream schemes.

9.2 The UK Government's **Heat and Buildings Strategy** will continue to monitor developments in geothermal heat and assess whether the technology provides a cost-effective option to help decarbonise heat.

9.3 As well as effectively preventing onshore wind (and fracking) through onerous local consent requirements, the government seems likely to prevent large scale solar farms by preventing installations on poor quality agricultural land designated as 3b.

PCC's CEAP initiatives

9.4 PCC has secured planning permission for a community solar farm (13MWp) at Chelson Meadow in partnership with Plymouth Energy Community (PEC). [152]

9.5 PCC will commence work to extend tidal flood defences at Arnold's Point along the Embankment up to the rail bridge.

9.6 PCC will test the yield from ground source wells in Millbay to provide low carbon heat. [153]

Non-combustion, zero carbon

9.7 Zero carbon renewables apply to energy sources that deliver electricity whilst in operation. These include **wind, solar, hydro, tidal, wave, and geothermal**.

9.8 Biomass involves combustion and is not zero carbon in operation. Biomass may be carbon neutral but only over an extended timescale whereas reductions are required this decade.

9.9 Hydrogen when used in a fuel cell is zero-carbon if it has been derived from zero carbon powered electrolysis. Fuel cells produce electricity by combining Hydrogen and Oxygen, leaving water as the only waste product.

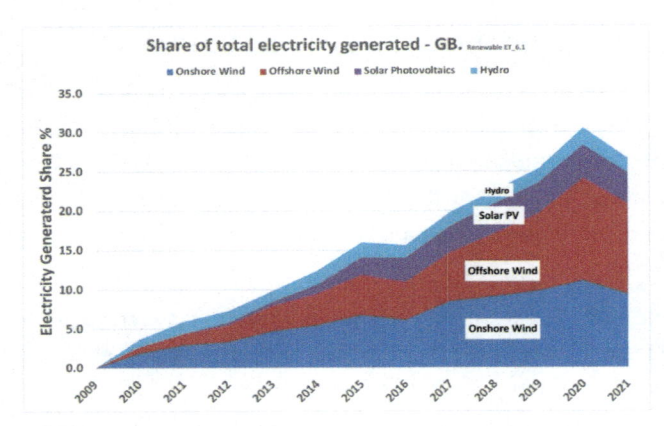

Share of total electricity generated - GB. Renewable ET_6.1

GB's zero-carbon renewables share of total electricity generated has steadily increased over the past decade. In 2020 it was around 30%.

Derating

9.10 The National Grid applies a derating factor to account for intermittency:

- Wind   0.430 (*Typically, 1MW installed = 3GWh per annum*)
- Solar photovoltaics   0.170. (*Typically, 1MW installed = 1GWh per annum*)
- Small scale hydro <5MW   0.365
- Tidal   (Not significant – no factor used)

9.11 Derating factors for intermittency of renewables are applied because renewables do not generate their maximum capacity continuously. Wind strength drops, night-time, dawn, and dusk all affect the strength of solar, rainfall affects hydro, tides affect tidal, etc. Applying these factors allows renewables to be accounted for fairly. Renewables when backed by energy storage can achieve much higher capacity factors

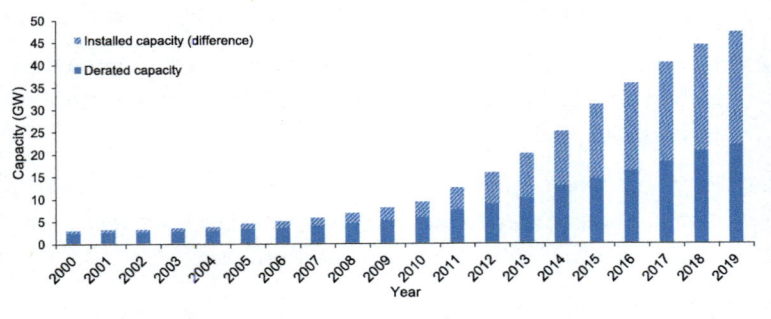

Derating effect on the installed capacity of renewables 2000 – 2019

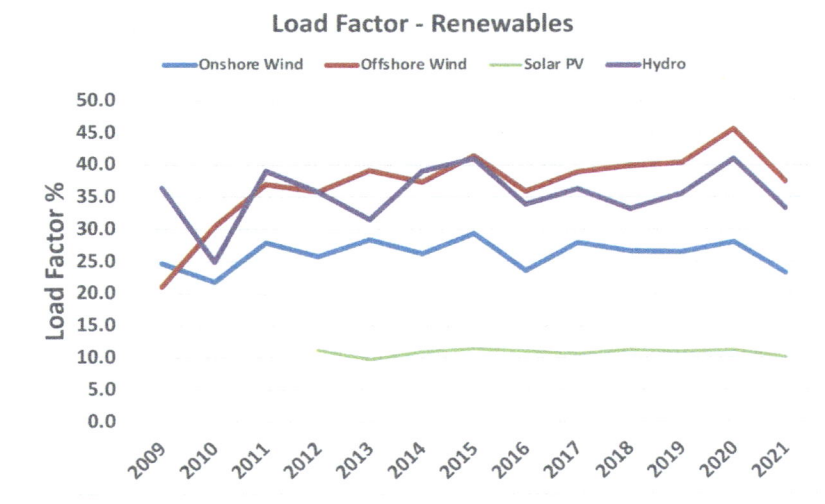

Combustion-Free Heat, Power, and Energy

Load Factor - Renewables

Load factor achieved by renewables – Load factor refers to the energy delivered compared to installed capacity.

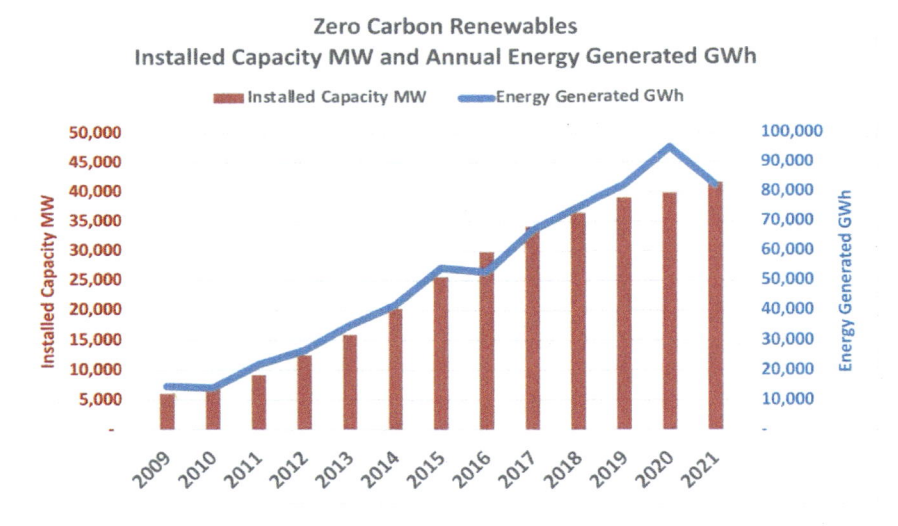

Zero Carbon Renewables
Installed Capacity MW and Annual Energy Generated GWh

Annual energy generated against installed capacity

9.12   Total zero carbon renewables of 1MW Installed delivers about 2- 2.5GWh generated annually!

Wind

9.13 Wind generated electricity has grown by 715% between 2009 and 2020. In 2020 wind generated 75,610GWh of electricity (offshore ~41,000GWh; onshore ~35,000GWh.) [154]

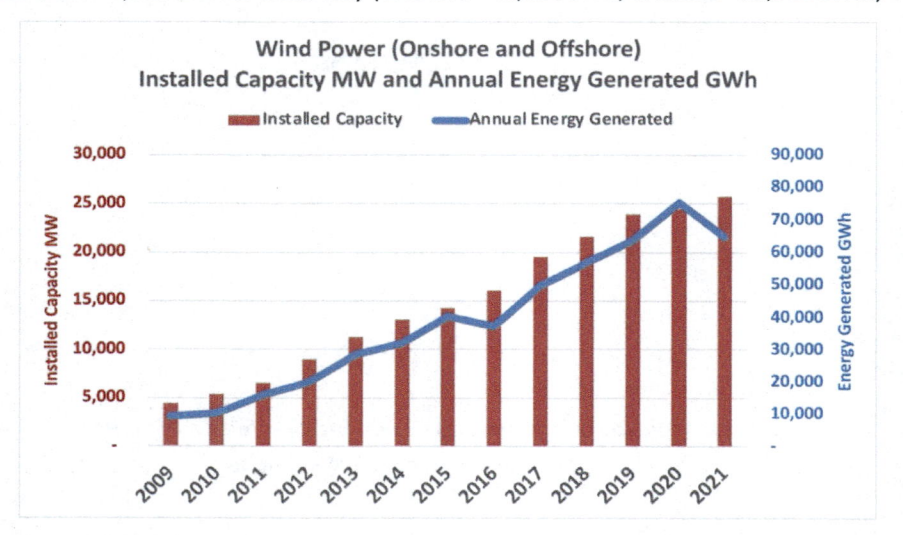

Energy generated by onshore and offshore wind power against installed capacity

9.14 Wind turbines are installed onshore and offshore. In the UK, Scotland has more onshore wind turbines and England has more offshore turbines

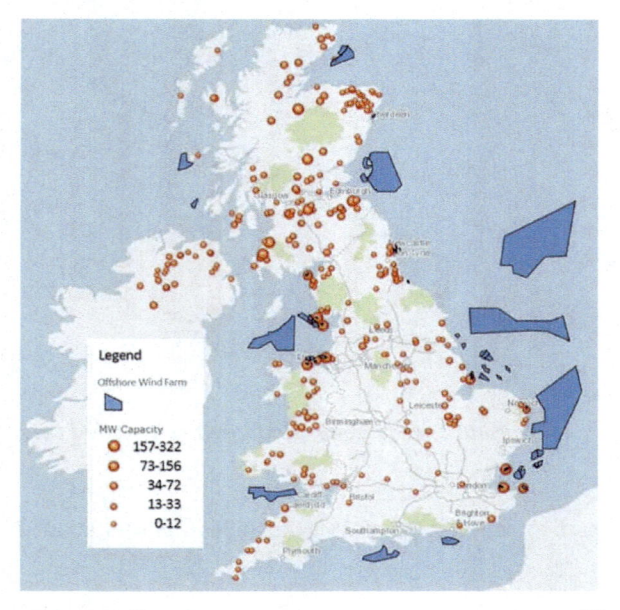

Wind turbine installations - Map from Renewable Energy Hub [155]

9.15 Onshore wind turbines in England are essentially subject to a moratorium. Any installation must fully address local concerns and gain unanimous approval. Scotland, Wales, and Northern Ireland have no such constraints. [156] The Onshore Renewable UK Industry Group are targeting 30GW of onshore wind by 2030. [157]

Installed Capacity GW Onshore (Proposed)	2021	2030
Scotland	8.4	20.4
England	3.1	3.6
Wales	1.3	3.6
Northern Ireland	1.4	2.5

9.16 The UK currently (2021) has 10.4GW of offshore wind operational; 3.6GW is under construction; 3.5GW in pre-construction phase; 9GW with consent authorised; 8GW with seabed rights authorised. 38.1GW total. [158]

9.17 The UK Government has a target of 40GW of installed offshore wind capacity by 2030. [159] Renewable UK research shows that the pipeline of offshore wind projects is 86GW of which 46GW are development projects not yet approved. This shows the scale of what is possible. [160]

9.18 Shallow water offshore wind platforms can rely upon a firm fixture to the sea floor. In deeper water, floating wind turbines adopt the technologies from the oil & gas industries – spar, semi subs and tension leg platforms.

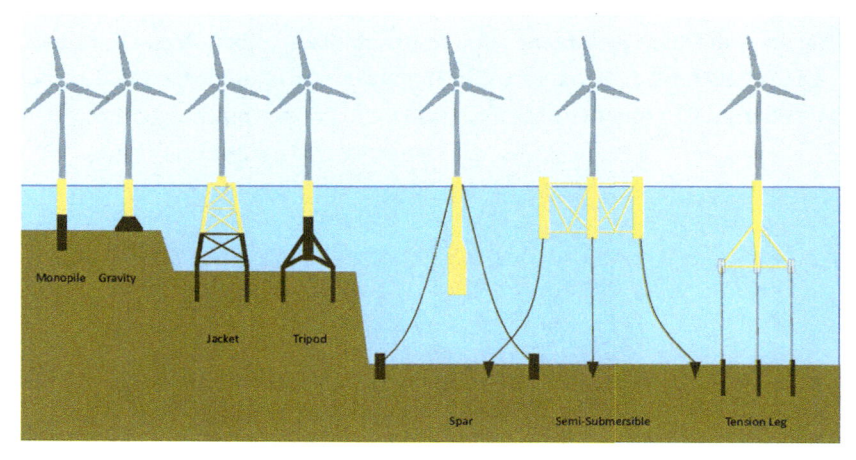

Technologies for deeper water wind turbines – Illustration CCA-4 [161] *From left to right – monopile, gravity, jacket, tripod, spar, semi-submersible, and tension leg.*

9.19 The ScotWind leasing round for offshore wind is expected to bring on a further 10GW.
[162 163]

ScotWind awarded offshore wind turbine sites

9.20 **Wind turbine technologies** have tended to coalesce around a single dominant design. This is a 3-bladed, horizontal axis (parallel to the wind direction), up-wind rotor, with pitch control for braking, and either an integrated gearbox or direct drive generator. Having more than four blades in a horizontal axis turbine leads to reduced efficiency because each blade is operating in the wake of the other blades. Tip speed is higher with the smaller number of blades, leading to greater noise signature.

GE Renewable Energy 14MW direct-drive generator, Haliade-X offshore wind turbine for Dogger Bank which will have 190 turbines in total. This has a 220m rotor and stands 260m high. Capacity factor is quoted as 60-64%. [164]

Combustion-Free Heat, Power, and Energy

9.21 Wind turbines are increasing in swept area and in hub height. In 1985, onshore wind turbines had a hub height of about 17m and delivered 75kW; by 2022, offshore wind turbines stood at 260m hub height with 220m rotor diameter delivering 14,000kW. Wind speeds tend to increase with increasing height above the ground/water although air density decreases, balancing out to some extent the wind strength. [165]

9.22 Horizontal turbines have a lower cut-in speed than vertical enhancing lower wind speed operation. Vertical axis wind turbines seem to dominate the domestic commodity market producing typically 600kW per turbine.

9.23 Rotor axis can also be vertical (perpendicular to the wind direction). The only large installation of vertical axis wind turbines was in California. With 500 turbines installed in 1986, giving a capacity of 95MW, annual energy generation in 1987 reached 105,000MWh (capacity factor of 0.12) before continual structural failures curtailed power generation. The installation was removed in 2005. [166]

Bladeless turbines

9.24 Designs for bladeless wind turbines are now being developed. The Vortex design uses oscillation (or resonant vibration) of the pillar to produce electricity. The Aeromine [167] design is particularly well suited to rooftop installations. Bladeless turbines bring the advantage of low visual impact, low noise and low running costs as well as being able to operate in very low wind speeds. [168]

9.25 Proposed developments from Vortex include a 1kW from a 9m high unit and 4kW from a 13m high unit. These designs have conditional support from the RSPB and CPRE (Campaign to Protect Rural England). [169]

Horizontal turbine and vertical axis wind turbines configurations, bladeless wind turbines and a Skysail prototype
(Photos - Martin Pearman CC BY-SA 2.0; Aeromine and Vortex; Skysail).

Skysail and other high altitude wind energy sources

9.26 Although the configurations for wind energy are largely settled, some alternatives remain available, albeit in prototype form. Skysail is one such system. A tethered kite is unreeled, generating power, and then reeled in by motor. The process is repeated to generate good baseload power. The prototype installation has completed its evaluation and is now generating power in Schleswig-Holstein.

Wind turbine environmental concerns

9.27 For onshore wind turbines: degradation of radar coverage / interference with wind farms [170]; bird strikes (seems to be greater with raptors and bats rather than with songbirds) [171]; visual impact to the landscape; shadow flicker, caused by the blade rotation on sunny days (flicker speeds are below those that can cause epilepsy).

9.28 The potential health impact of [onshore] wind turbines (Chief Medical Officer Ontario, Canada) review concludes on infrasound that while some people living near wind turbines report symptoms such as dizziness, headaches, and sleep disturbance, the scientific evidence available to date does not demonstrate a direct causal link between wind turbine noise and adverse health effects. The sound level from wind turbines at common residential setbacks is not sufficient to cause hearing impairment or other direct health effects, although some people may find it annoying. [172]

9.29 For offshore wind turbines, environmental concerns include: [173 174 175 176]:
- fish and marine mammals including sea turtles. Changes to pelagic and benthic habitats particularly from construction and decommissioning, benefits from the increase in shellfish attaching to structural foundations and an artificial marine reserve sheltered from fishing.
- birds, particularly at peak migration times, and bats. Whilst the effects for onshore or nearshore are better understood, offshore is less well studied. Many seabirds fly at altitudes below the wind turbines but where sea birds fly at wind turbine altitudes there is evidence of avoidance action being taken. Bats suffer from lung collapse due to the low pressure from the rotating blades. Mitigating action for these issues include avoiding low-wind speed times and significant migrating times.
- noise, particularly from piledriving during near shore construction and decommissioning – (the presumption is removal). Pile driving (100-500Hz) and air gun seismic surveys (10-120Hz) produce extremely high intensity noise and rapid intervals. Some mitigation is possible by starting the process with lower-level noise and longer intervals – there is some evidence that fish and marine mammals avoid the noisy areas if this is done but the process must be slow as high intensity noise has been measured at distances or 70km and more. The noise issue is not considered a concern for tethered floating wind farms suitable for deeper water.

Noise during operation and the effects on breeding and feeding are not well understood.

- Electro Magnetic Fields EMF – has been reported as an issue with brown crabs. The EMF from the transmission lines feeding power from offshore wind farms seems to cause the crabs to stop moving or feeding. [177]

9.30 Blades are manufactured from glass and or carbon fibre encased in a plastic resin making them difficult to recycle. One approach is to chop them up and use them in decking boards or similar [178] or even to turn them into 'gummy bear' sweets [179].

Solar Photo Voltaic (PV)

9.31 **Photo Voltaic Solar** converts sunlight into DC (Direct Current) electricity. To use this electricity within a household an inverter is required to produce AC (Alternating Current) electricity.

9.32 Solar panels deliver power during daylight hours; the stronger the light (not heat), the greater the power produced. As well as daily variability, power produced varies during the year due to the variation in light intensity and angle of the sun. Solar PV is used in small domestic rooftop installations, typically 4kWp (peak), or in large solar farms, beyond 70MW with approved proposals for installations up to 350MW. [180]

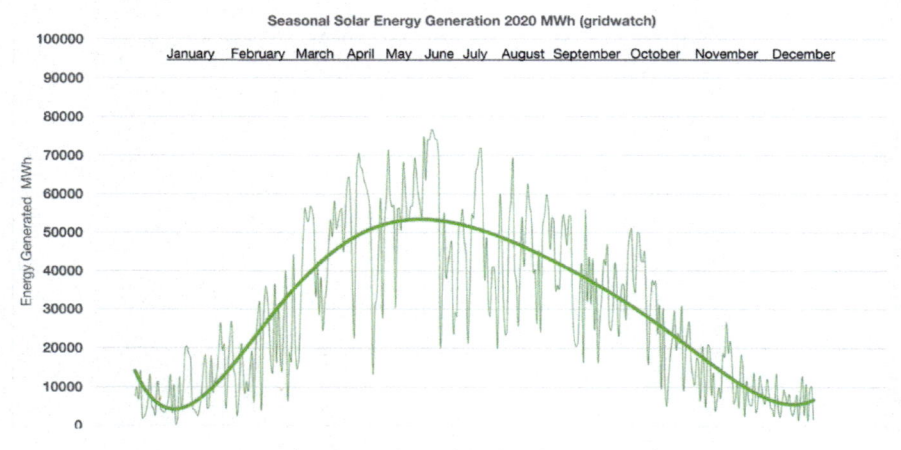

Solar energy generation is seasonal

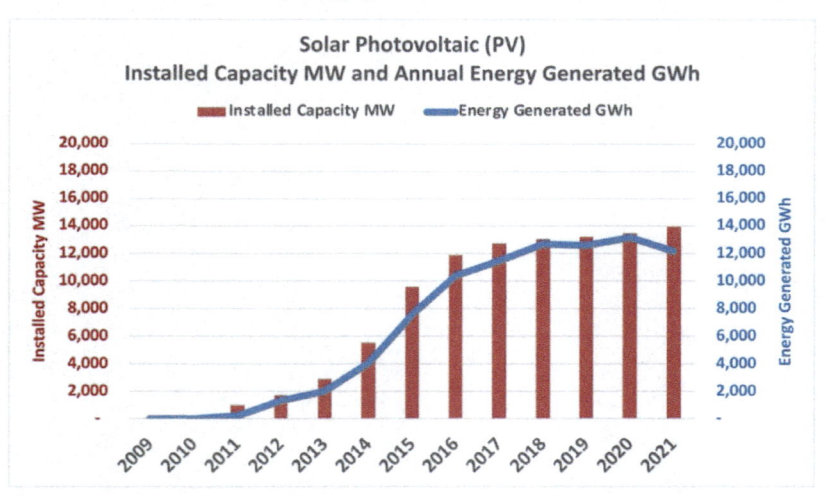

Solar power growth has been significantly affected by subsidies. By 2021, there was 14GW of Solar generating 12,000GWh per annum. The Government plans for 70GW of solar by 2030, generating around 60,000GWh per annum.

Combustion-Free Heat, Power, and Energy

9.33 The UK's largest **solar farm** will be the 350MW Cleve Hill Solar Park in Kent. It is expected to produce 264GWh annually, lowering CO_2 emissions by 68,000 tonnes a year. Around 57% of the total solar capacity relates to large ground-mounted installations. [181]

Solar Farm at Ernesettle, Plymouth (Photo Malcolm Teague)

9.34 Most solar installations in the UK are installed to deliver the highest peak power, meaning south facing panels at the optimum elevation angle. However, to extend the generation time and better cover the seasons, east-west installations are now being installed and are being backed by Battery Energy Storage Systems BESS. [182]

9.35 Typical **solar panel efficiency** is between 21 and 22.5%. This efficiency affects the size of the panel as it is a ratio of sunlight received versus power delivered. Unlike fossil fuel power generation, this is not really a ratio that affects the cost of use although it will impact installation costs. Panel sizes have increased over the past decade. This is particularly beneficial for the installation costs of large solar farms. [183]

9.36 Solar PV sites can easily be returned to their original, or better, state at the end of their lives. National planning policy in England and Wales requires a solar farm development to deliver a biodiversity net gain and greater biodiversity resilience. [184 185] Solar Farm sites can be a haven for biodiversity. [186], can be integrated into agricultural settings, shielding plants growing beneath them from extreme weather whilst allowing sheep to graze beneath them. [187 188 189]

9.37 The National Planning Policy Framework expresses a preference for large scale ground mounted solar PV systems to be sited on agricultural land graded 3b, 4 and 5. [190] DEFRA has recently indicated it is considering a solar ban on lower quality English farmland. [191]

9.38 Floating solar farms can also be installed on reservoirs. An analysis of the UK's top ten reservoirs indicates the potential for up to 6.8GWp, about half of the 2022 cumulative installation. Potential concerns include seasonal fluctuation of water level affecting the number of panels that can be installed; grid connection availability; and higher upfront and maintenance costs. [192]

9.39 For Plymouth, planning permission for a community owned solar farm on the old landfill site at Chelson Meadows has now been granted. The proposal is for a 13.2MW solar array backed by energy storage. The solar farm should generate enough energy to power 3,800 homes (~3.3% of Plymouth's housing stock). The scheme is supported by the Rural Community Energy Fund Programme. [193]

9.40 A recent innovation has been to site solar farms offshore in and around wind farms thereby improving the power output for a given area. For countries such as the Netherlands, land space is very limited whilst sea space is comparatively open. Tests have proven the ability for **floating solar** to withstand the rigours of the offshore environment. [194 195]

9.41 **Domestic rooftop solar** can be incorporated on build or as a retrofit.

Environmental impact of solar [196 197]

9.42 The environmental impact of solar PV during operation is minimal in comparison with other technologies for producing energy. Panels are easily removed if no longer required.

9.43 When considering the manufacture and disposal phases, hazardous material, water resource pollution, and emissions of air pollutants during manufacture should be considered. Using less material per kWh and improving the carbon emissions of the country of manufacture's electricity grid will lead to improvements. The key issues concern quartz mining and refining silicon which entails the use of silicon tetrachloride. Some manufacturers of solar PV recycle the silicon tetrachloride and some discard it. When exposed to water, silicon tetrachloride produces hydrochloric acid, acidifying the soil and producing toxic fumes. Hydrofluoric and sulfuric acids are used when cleaning silicon wafers.

9.44 Solar farms do take a lot of land area (~half a square kilometre (10 acres) per MWp) but done sympathetically, biodiversity does not seem to be negatively affected. Clarkson & Woods (ecological consultants) in research conducted in 2015 concluded: solar farms had greater botanical diversity, greater invertebrate abundance, greater diversity of birds - and that solar farms are of particular benefit for bird conservation. [198 199]. Bat activity may be lower within ground mounted solar arrays. Bats are also affected by in-roof mounted solar panels as they change the background temperature of the roof. [200]

9.45 Professional and independent consultants, BSC-Ecology, having done a literature review in 2019 on the *"Potential ecological impact of ground-mounted photovoltaic solar panels'* concluded: *"From the body of research reviewed, it is likely that the majority of concerns that have been discussed in the media are not well founded, or are based on scientific experiments that were not specifically designed to evaluate ecological impacts of ground mounted solar PV sites".* [201]

Hydro

9.46 Hydro energy generation derives power from a dam, a weir or from flowing water. Turbines generate power from the head of water or from the water flow.

9.47 There are three types of hydro power [202]:

- **Storage** involves a single dam and reservoir
- **Pumped storage** involving two reservoirs with water being pumped to the higher reservoir and generating power as it flows back to the lower reservoir.
- **Run-of-river** uses the natural flow of a river and a weir, diverting water through a powerhouse containing a turbine generator.

9.48 The UK's hydro energy generation is mostly from small sites based in Scotland. Small scale hydro is <5MW; Large scale hydro >5MW. Micro scale hydro is <50kW. There is little scope to expand hydro much further; sites are available but wind and solar generate a higher return on investment.

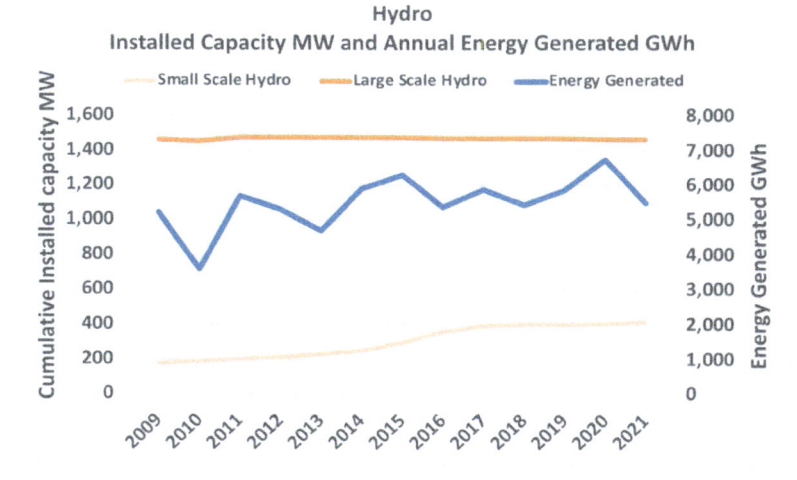

Electricity generated from hydro over the past ten years has remained steady

9.49 Hydro contribution to UK electricity generation was 1.3% in 2020. It is a complementary energy source to other intermittent renewables. [203]

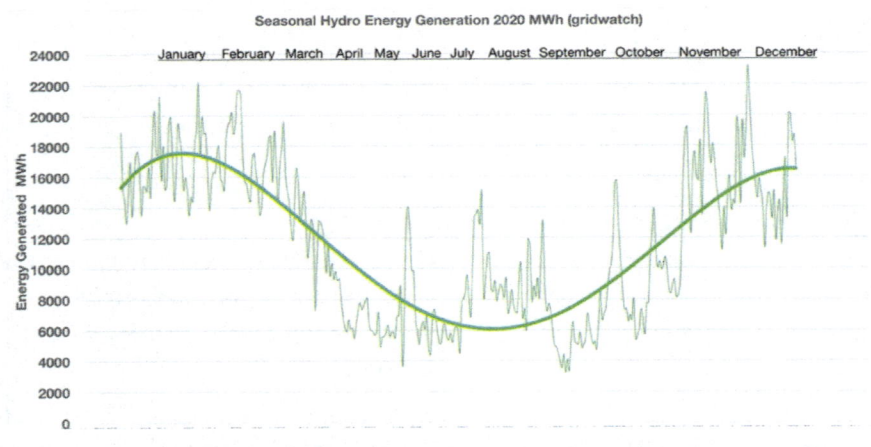

Climate Action Plymouth – Sustainable Power and Energy Group

Seasonal Hydro Energy Generation 2020 MWh (gridwatch)

Hydro seasonal variability - there is more rain in winter than summer.

Severn Estuary Tidal Barrage

9.50 After many false starts, on the 8th of March 2022, the Government relaunched studies into harnessing the hydro power potential from the large tidal range of the Severn Estuary. Potentially, a Severn tidal barrage could meet up to 7% of the UK's electricity requirements. Any barrage scheme would likely take more than ten years to build and cost more than £25bn. [204]

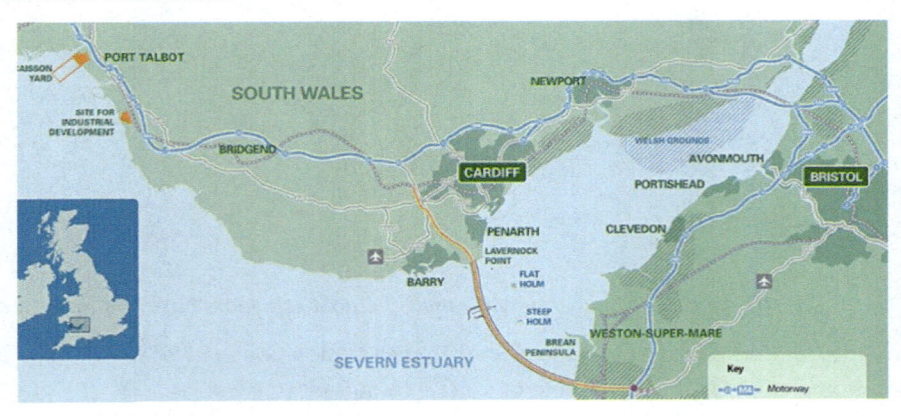

Possible Severn Tidal Barrage Scheme (The Greenage [205])

9.51 Tidal lagoons as well as a barrage are under consideration. Electricity would be predictable but variable due to the timing of tides that may often be out of phase with peak power demands. Some of this could be mitigated by energy storage.

Environmental impact of hydro [206 207 208]

9.52 Hydro appears a very clean and environmentally friendly source of energy and generally they are net positive. However, there are impacts on wildlife habitats and on migration paths. Damming rivers has a considerable negative impact on rivers and their wildlife.

9.53 Dammed rivers and reservoirs can contribute to increased greenhouse gases from decaying trapped vegetation.

9.54 Generally, most negative impacts can be mitigated to some extent. Silt and sediment build up seems unavoidable, the consequences of which may take many decades to manifest.

9.55 Power turbines kill around 2-5% of the fish that flow through them.

9.56 Reservoir construction can also have a social and economic impact on those that live in or around the facility. Livelihoods can be detrimentally affected.

Tidal stream

9.58 In November 2021, the UK Government announced an investment of £20m per year ringfenced for tidal stream electricity. [209]

9.59 In 2018, 22 tidal stream technology developers were active in the UK. Projects in place or under construction are at Meygen site at Pentland Firth and Nova in Shetland. Meygen Phase 1A is installed at 6MW. Meygen 1B adds a further 4MW and Phase 1C adds 73.5MW. Increases in power up to 398MW are planned. [210]

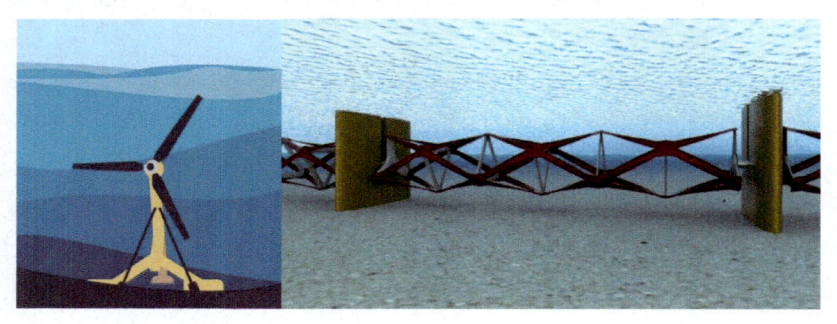

Typical axial (CCA-4[211]) and horizontal tidal turbines (Keplerenergy)

9.60 **Nova Innovation** have a successful installation in Bluemull Sound, Shetland. Initial installation was for three two-bladed 100kW direct-drive turbines for a capacity of 300kW.

9.61 The **Bluemull Sound installation** has now been extended to 600kW and connected to a Tesla **Battery Energy Storage System** which ensures that the power station can deliver constant, steady-state power to meet consumer demand. [212]

9.62 **Orbital Marine Power** have been awarded two CfDs delivering 7.2MW as part of the Government's renewable energy auction. [213]

Orbital Marine Tidal Power

9.63 Existing underwater tidal schemes adopt axial flow turbine blade genèrators. Alternatives developed by the University of Oxford, the **Kepler Transverse Horizontal Axis Water Turbine,** provides an alternative technology.

9.64 Given that tides ebb and flow with quiescent phases in between, energy generated will be intermittent although entirely predictable. Energy produced will be in phase with demand for some of the time and out of phase at other times. Tidal is therefore probably best connected through BESS (as Nova have done) or used for hydrogen production.

9.65 As with wind energy arrays, tidal arrays are subject to interactions and interference between turbines, with upstream turbines reducing the amount of power developed by downstream turbines. The EnFait project is studying this phenomenon to establish what degree of optimisation is possible. [214]

Environmental impact of tidal stream [215]

9.66 Tidal stream is in its infancy and little serious study has been done on the environmental and ecological impact of such installations.

- Construction will have a similar impact on fish and sea mammals as piling for wind turbines. These impacts will generally be short term.
- Concerns have been raised about the impact on currents and waves, impact on substrates and sediments, habitats for benthic (seabed) organisms, noise from turbine motors or gears, cavitation noise from rotors, electromagnetic effects, toxicity of paints, lubricants and antifouling coatings, and fish and mammal movements and strikes by rotors.

Wave Power

9.67 Wave power harvests the energy from the waves that are produced by winds and storms. The distance between the wave peak and the wave trough (wave amplitude) together with how quickly the peak becomes a trough and then a peak again (wave period) indicates the approximate amount of energy that can be produced. Wave power has had a chequered history in the UK. [216 217]

9.68 The method for harvesting the energy is of one or two types regardless of whether the wave energy device is offshore, nearshore, or on the shoreline. Either the device moves hydraulic rams that are then used to generate electricity or the rise and fall of the waves drive a turbine directly to produce electricity. Wave power is an intermittent renewable. It is predictable in the same way as wind or solar.

9.69 The **Salter Duck** was a device that harvested energy by facing the waves with the nodding mechanisms driving hydraulic rams to generate electricity. It was proposed in response to the 1970s energy crisis, The device was never tested in realistic wave conditions – Salter believes the Duck was derailed following pressure from the nuclear industry. [218]

9.70 Salter has subsequently been involved with the Pelamis Sea-Snake which has now gone into administration. The Sea-Snake lay in line with the wave direction – the articulating device following the waves with hydraulic rams used to produce electricity. The device will now be disposed of and sold for scrap. [219]

9.71 Scotland (Wave Energy Scotland), together with a regional Basque Country authority (Basque Energy Agency), have allocated £20m of funds to seven promising wave power projects commencing 2022. This is match funded by the EU. [220] Following completion of the first stage, the five most promising technologies will be selected. The final phase will select three devices for testing in real conditions in 2025. Given the many years of failure of wave power technologies, this is a sensibly cautious approach to what should be an important renewable energy technology for the UK.

Pelamis Sea-Snake Wave Power Generator (CCA-3 [221]) and Mocean Energy Blue X Wave Power hinged-raft machine [222]

9.72 Cornwall's wave hub has been sold to a wind power generator as there have been no successful wave projects able to take advantage of the facility. [223]

9.73 The above projects have aimed to extract energy from offshore. **Eco-Wave Power** take a different approach by aiming to extract power from shoreline and near shore installations. [224 225] Although the amount of energy available from offshore is potentially greater, it is often confused rather than directional as well as being more expensive to bring ashore.

9.74 Levelised Cost of Electricity (LCOE) for a commercial scale Eco-wave installation (~£35/MWh) is targeted at or below those of offshore and onshore wind as well as below that of large solar farms. By utilising breakwaters to install the wave power equipment, installation costs, risks, and ongoing maintenance are all predicted to be lower than offshore installations.

Eco Wave power installation off a breakwater (Photo Elarbiser)

9.75 Projects are underway in Israel and in Gibraltar. Plymouth has the benefit of two large breakwaters, one at the entrance to The Sound (1,560 metres) and one at Mount Batten (279 metres). Both may be suitable for such an installation although a tidal range between high and low tide of 5.9m may pose problems – Eco-wave declined to comment.

Environmental impact of wave power [226]

9.76 Despite being many decades in development, Wave Power is still in its infancy and, as such, there is little real data to draw on when considering its environmental impact. Concerns have been raised about coastal erosion, sedimentary flows, exclusion zones for fishing and recreation, navigation hazards, and operational noise pollution. There are also the same issues as other marine projects of electromagnetic fields, construction noise, hydraulic oil leakage, and paint or anti-fouling effects on the marine eco-system.

Geothermal

9.77 Geothermal energy includes the low-grade heat stored just below the ground (<200m) and which can be exploited by heat pumps, and deep geothermal energy for heat at >500m. Deep geothermal energy derives from the heat produced when the Earth was formed together with the heat from the decay of radio-active elements within the Earth's core. [227]

9.78 Even at depths of 5000m, temperatures are only in the region of 139°C and so, whilst insufficient for a steam turbine, this heat can be used for **direct heating of houses** including district heating, provided they are within a reasonable distance of the drilling.

9.79 In areas where there are radiogenic granites at economically drillable depths then much higher temperature water can be extracted sufficient for steam power generation. However, granite is not very porous and hence **Engineered Geothermal System EGS** technology is necessary utilising naturally occurring fractures in hot dry rock. [228]

9.80 A borehole study concluded that sufficient heat >200°C existed in radiogenic granite below Bodmin Moor, Lands End, Carnmenellis, and St Austell, and 185°C in Dartmoor. The study concluded that Camborne, Penzance, St Austell, Redruth, and St Ives offered the most likely conurbations for geothermal district heating. [229]

9.81 A further study in 2017 by Busby and Terrington concluded that at drill depths of 4-5km, the potential for deep geothermal energy is restricted to the granite areas of Southwest England, Northern England, and the East Grampians. Newer drilling technology may allow deeper drills up to 7km enabling deep geothermal energy for a much larger part of the UK. [230]. The US Department of Energy DoE is funding projects to develop faster geothermal drilling techniques. [231]

9.82 The only **existing geothermal scheme** is a district heating project in Southampton. The system provides 13GWh of heat to several buildings including the West Quay shopping centre, the civic centre, and the hospital. [232] The well depth is 1,800m and the temperature is 76°C.

9.83 The **Eden Project** [233], near St Austell in Cornwall, is aiming to develop a Direct Use Hydrothermal System and has started test drilling. The borehole has been sunk to a depth of 4,871m and has found its target fault structure. A second well will now be sunk, close to the first and to a similar depth. This should provide the opportunity for an Engineered Geothermal System generating electricity. The two-well Eden Geothermal project aims to supply heat to 35,000 homes and electricity to 14,000 homes. The project suffered seismic activity (1.6 magnitude) during testing of the rock permeability phase in March 2022. [234 235] Operations have now been restarted without any further felt seismicity. [236]

Combustion-Free Heat, Power, and Energy

9.84 The **United Downs Deep Geothermal Power Project** UDDGPP [237] uses the hot granite rocks near Redruth, Cornwall. Two deep wells have been successfully drilled, one to a depth of 5275m and the other to a depth of 2,395m. The expected water temperature is 190°C. The project, by Geothermal Engineering Ltd GEL, is to demonstrate the feasibility and to produce around 1-3MW of electricity. GEL plans to deliver four deep wells in Cornwall producing 20MW of baseload electricity and 100MW of heat. Over the next 20 years, GELs target is 500MW of power. A sister project at Jubilee Pool is being used to directly heat a swimming pool. [233]

9.85 Britain sits on a considerable number of **disused deep coal, tin or other mines**. Flooding these and extracting the heat could provide a useful and carbon-free heat source for district heating schemes. Such schemes are under development in Seaham Garden Village, Durham, in Hebburn, in Gateshead, and in Nottingham. The Government has provided funding to explore the potential of disused tin mines in Cornwall as a similar source. The warm water in the mines is used together with heat pumps to bring it to a temperature suitable for home heating. [239]

9.86 There are some indications of mining for minerals in and around the Plymouth area. A string of mines including Wheal Southway, Wheal Looseleigh, Wheal Whitleigh and Wheal Genny were worked between 1850 and 1857. Mine workings were later believed remediated by the council. [240]

Environmental impact of geothermal power [241 242 243]

9.87 Geothermal energy is not really a 'renewable energy' but can be considered 'sustainable' due the large amounts of heat available from radioactive decay within the earth's core.

9.88 Many of the environmental issues with geothermal are common with natural gas hydraulic fracking – water contamination, water usage, and a greater earthquake frequency

9.89 Air-emission of greenhouse gases are about a tenth of natural gas. Greenhouse gases include carbon dioxide, ammonia, and methane. Air emissions also include boron and hydrogen sulphide/sulphur dioxide. Using exhaust gas washing systems (scrubbers) produces a hazardous waste sludge of sulphur, vanadium, chlorides, arsenic, mercury, nickel, and other heavy metals.

9.90 Geothermal sites can run out of heat and require new wells to be drilled. The timescales will vary.

10. Nuclear-Fuelled Steam Power Generation

No new nuclear power generation plant has been commissioned in the UK since 1995. UK has no indigenous large nuclear power plant design capability. Nuclear is supported by both major political parties but not by all political parties.

Large nuclear plants have long timescales for build, are very costly, and are commercially highly risky.

Electricity costs, based on Hinkley C, from nuclear appeared very high in 2020 although the UK Government's predicted LCOE for the next plant is more acceptable, particularly given the high fossil fuel generated electricity costs.

Major questions remain about the safety of large nuclear plant (peacetime and wartime) and the disposal of spent fuel.

Large nuclear power plants do not complement zero-emission renewables.

Small Modular Reactors SMR have the potential to contribute more quickly than large nuclear plants but are currently only in the study phase. They share many of the downsides of large nuclear plant.

Advanced Modular Reactors AMR, such as the Moltex Stable Salt Reactor, has the potential to provide inherently safe, low-cost, low-carbon, power plant that uses existing nuclear waste as a fuel. SSR's load following capability means they are complementary to renewables. However, they are unproven and some years away.

UK Government strategy and initiatives

10.1 The UK Government in its **British Energy Security Strategy (BESS)** projects up to 24GW of nuclear energy by 2050 – up to 25% of Great Britain's total electricity demand.

10.2 The **Green Industrial Revolution (GIR)** proposes further investment in Small Modular Reactors (SMR) and Advanced Modular Reactors (AMR) and to build a demonstrator (SMR or AMR) by early 2030.

Nuclear power plant in UK

10.3 The UK opened the world's first nuclear power station in 1956 at Calder Hall in Cumbria. The design used a British designed Magnox Reactor, optimised to use natural uranium. Carbon Dioxide was used as the coolant. Calder Hall and Chapelcross reactors were designed to produce plutonium as well as electricity. Ten further Magnox type reactors were opened over the next 15 years, all designed to maximise electricity generation. [244] All Magnox reactors are now closed.

10.4 The next seven were Advanced Gas-cooled Reactors (AGR), a British design using carbon dioxide as a coolant producing approx. 1200MWe of power. An AGR requires enriched uranium and produces a higher steam temperature than Pressurised Water Reactors (PWRs). Four AGRs have now closed with the other three due to close by 2030. [245] These are Heysham1 & Hartlepool in 2024, and Torness in 2030.

10.5 The most recent nuclear power station, Sizewell B, a Pressurised Water Reactor (PWR) built to a Westinghouse/Bechtel, USA/French design, opened in the UK in 1995. It is scheduled to continue in operation until 2035 delivering approx. 1200MWe of power. [246]

10.6 MPs have called for decommissioning delays for UK's ageing nuclear fleet. The Department of BEIS and EDF will be tasked to double check the feasibility of extending the lives of the remaining nuclear stations to ensure energy security and ease energy prices. [247]

10.7 The remaining operational nuclear power stations are all AGR plus the one PWR. Total capacity (2022) is 8GW.

10.8 Hinkley Point C1 and C2 are under construction. Each reactor will produce up to 1.63 GW. They are a European Pressurised Reactor (EPR) to a French EDF pressurised water design. Two EPR reactors are operational in China. Finland's EPR reached criticality at the end of 2021, has begun operation March 2022.

10.9 Hinkley Point C is now ten years late having been rescheduled to June 2027 from the original 2017 operational date. [248 249]. The Flamanville EPR in France, which Hinkley Point C is based upon, has faced numerous quality issues and delays: originally expected to begin operating commercially in 2013, the fuel loading date is now postponed to the second quarter of 2023 with a scheduled operational date of 2024. [250]. Hinkley C contains two reactors, total power 3.2GWe.

10.10 The strike price (for Contracts for Difference purposes) for Hinkley Point C is £89.50/MWh (2012 prices) [251] which was more than double the price of electricity at the time of signing (£45-55/MWh) but very competitive with 2022 energy crisis day-ahead spot prices for electricity of around £330/MWh [252]. Hinkley C has a funded decommissioning programme (FDP) as part of its contract ensuring that taxpayers do not bear the burden of the costs. [253]

10.11 The UK Government's next scheduled large reactor is Sizewell C. This will be built under a Regulated Asset Base RAB model whereby some of the costs and risks are transferred to the Government and the contractor, EDF, in return for reducing the cost of the project. Sizewell C has now received 'project development consent' to proceed and will be to an EDF EPR design. [254] The decision is being appealed because of insufficient cooling water in a water stressed area. [255]

10.12 All nuclear power plants use 'fission' to produce heat that drives a turbine generator to produce electricity. The diagrams below illustrate a pressurised water reactor PWR/EPR, a similar design for Hinkley C and Sizewell C.

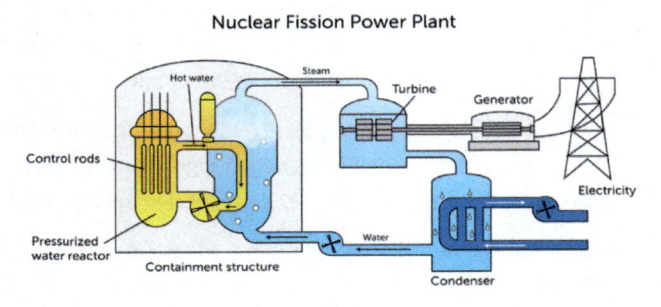

Pressurised Water Reactor

10.13 Fission involves splitting the atoms of uranium (or plutonium) and in the process releasing heat. The heat is used to raise steam which drives a turbine to produce electricity

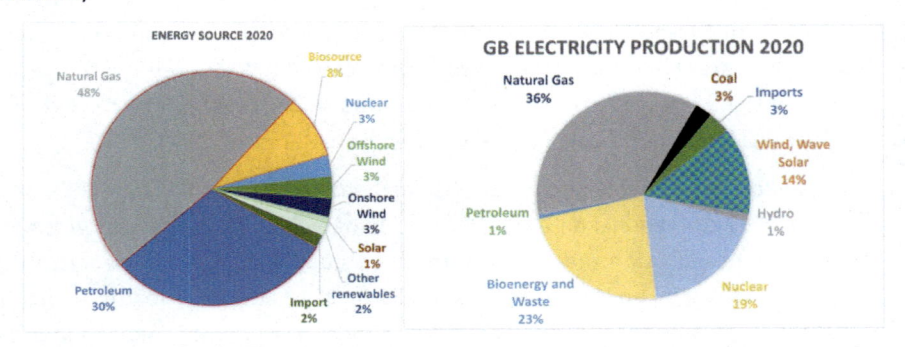

Nuclear power contributes just 3% to the total UK energy needs (2020) but 19% towards electricity production (2020) (DUKES) [256].

Large-reactor capability

10.14 Despite leading the world in building nuclear power plant, the UK no longer has a **large civil nuclear reactor design capability**. [257] The last UK reactor to be built was in 1995 to a US Westinghouse PWR design. The plant is now owned and operated by

EDF. Hinkley Point C is built to a French European Pressurised Reactor (EPR) design, again owned, and operated by EDF.

10.15 AS of May 2022, 441 nuclear reactors operate in 32 countries producing 393GW or about 10% of the world's electricity demand. Fifty-three (53) reactors are currently under construction, mostly in Asia and Russia. A further 96 are planned with 325 proposed. [258]

10.16 Nuclear power plant designs are available from Canada (CANDU), China (HTR-PM/AP1000/CAP1400/Haulong-2), France (EPR), Japan, South Korea, and Russia (VVER-1000).

10.17 Many countries worldwide have now committed to halting the decline in nuclear power and intend to restart building or to increase building of more nuclear power plants. Addressing climate change and securing energy sovereignty are cited as the reasons. France is to build 14 new nuclear reactors to an optimised EPR2 design. [259] South Korea considers nuclear to be the main source of electricity over the next 60 years. [260] The UK plans eight new nuclear reactors and is considering extending the life of existing ones. [261]

Small Modular Reactors (SMR)

10.18 Building large nuclear power plants is a very long and very expensive business. Risks are far greater than any commercial organisation can withstand, leading to direct or indirect government investment and risk sharing.

Reactor Core and SMR nuclear power plant *(Rolls-Royce)*

10.19 Proposals exist in many countries for a 'factory' produced small modular reactor that can be series built in a factory and transported to a site, plugged in, and become operational, enhancing the number of available sites. In the USA, "The first generation-IV nuclear (small modular) reactor design has been approved for certification by the US Nuclear Regulatory Commission". [262]

10.20 In the UK, Rolls-Royce has developed an SMR where the core can be built in factories, such as their factory in Derby, and then be delivered to the balance of plant at the site. The site is expected to be about the size of two football pitches. [263] The SMR is designed to deliver approximately 470MWe on completion.

10.21 The Rolls-Royce SMR will not be designed to load follow and so **will not be complementary to renewable intermittency.**

10.22 The Rolls-Royce SMR is based on their experience of designing and delivering submarine reactor technology (PWR) but no defence technology is carried over into the SMR, say Rolls-Royce [264].

10.23 Once designed, an SMR is estimated to take just four years from order to power production with the first unit being on-grid by 2029, assuming an order by 2024 [265]. Up to 16 power plants are planned to be operational by the mid 2030s [266]. Refuelling is expected every two years.

10.24 A recent study published in the Proceedings of the National Academy of Sciences in May 2022 show that waste from SMR's (Pressurised Water, Molten Salt, and Sodium cooled) will increase the volume of nuclear waste in need of management and disposal by between 2 and 30 times. In addition, due to increased neutron leakage, SMR's will generate at least nine times more neutron-activated steel than conventional nuclear power plants. [267] This is a significant increase in the amount of nuclear waste to be managed.

10.25 The Rolls-Royce SMR utilises the current UK supply chain (80%), with only the steam generating plant being sourced from overseas. Funding has been provided by the UK Government [268]. SMRs can be plugged in at any of the decommissioned nuclear power plant sites thereby minimising approvals and maximising acceptance by the local community [269].

Advanced Modular Reactors (AMR)

10.26 The UK Dept of Business, Energy, and Industrial Strategy (BEIS) has funded a two-phase programme for Advanced Nuclear Technologies. Phase 1 provides up to £4m, with contracts between £300k and £400k. Phase 2 provides up to £40m for selected projects from phase 1. [270]

10.27 Advanced Nuclear Technologies aim to demonstrate reactors that are smaller than traditional reactors, designed so that they can be factory constructed to reduce costs and build risks. Advanced Modular Reactors are expected to have a higher temperature output, perhaps 700-950°C compared to Light Water Reactors (LWR) at 300°C or Advanced Gas Cooled (AGR) at 600°C.

10.28 The winners of Phase 1 include a few UK firms amongst those shortlisted for AMR include: Advanced Reactor Concepts LLC, DBD Ltd, Blykalla Reaktorer Stockholm AB, **Moltex**, Energy, U-Battery Development Ltd, Ultra Safe Nuclear Corporation, and Westinghouse Electric Company.

10.29 The Moltex **Stable Salt Reactor (SSR)** [271] uses a rejected idea for an aircraft nuclear power plant conceptualised in the sixties. The concept did not work in an aeroplane

Combustion-Free Heat, Power, and Energy

because the varying levels of gravity experienced by an aeroplane affected the cooling capability of the salt. Moltex have developed and patented the technology for use on a static ground station. Stable Salt Reactors appear in the BBC's '39 ways to save the planet'. [272] Innovate UK supported Moltex Energy's **Stable Salt Reactor SSR** in 2015. [273]

10.30 Moltex SSR design recycles and consumes the spent reactor fuel, contributing towards reducing nuclear waste stockpiles whilst generating zero carbon, zero emissions, energy. SSR's have good load following capabilities and are hence complementary to intermittent renewables. The SSR's design has overcome corrosion and engineering complexities from other Molten Salt Reactor MSR types. LCOE of the SSR is predicted to be quite low at ~ £37/MWh.

10.31 **SSR's are inherently safe.** The molten salt fuel is contained within a conventional nuclear fuel tube which is in turn cooled by a separate molten salt that transfers heat out of the reactor to produce steam to generate power. Separating the molten salt fuel from the molten salt coolant means that, even in the unlikely event of a fuel tube leak, critical mass cannot be achieved and hence no explosion risk exists. [274]

10.32 **SSR's are fuelled by existing nuclear waste.** The SSR is a fast-spectrum reactor. It burns the longer lived heavier nuclear waste quicker, reducing it to a radioactive product more like mined uranium, something that is much more manageable. [275 276]

10.33 Moltex have received US$4.5m from the US and a significant Can$50.5m from Canada for research on its SSR SMR. [277] A demonstration SSR is planned for 2030 at New Brunswick in Canada. [278] The full-size reactor would be 1GWe with the prototype running at around 150MW. Its non-pumped, fuel-in-tube, design is inherently safe and cannot boil in the event of an unplanned shutdown.

Nuclear energy (fission) security, environmental risks, and concerns

10.34 Nuclear fission power plants offer a reliable, carbon-free electricity during operation, take up a small land footprint for the high-power output, but have a significant public acceptance difficulty. These difficulties concern: availability and security of fuel supply, operational safety, waste disposal, and weapon proliferation.

10.35 Uranium is the dominant fuel used in fission reactors. **Uranium is finite and is not renewable**: there is enough uranium to fuel today's reactor output for a further 90 years at ore prices up to about three times today's cost. The IPCC states there is enough uranium for a further 130 years at today's rate of use or 250 years if all uranium resources are used, although costs of extraction would be higher.

10.36 Thorium can also be used in some reactors and is about three times more abundant than uranium. It is difficult to weaponise thorium fuel or its byproducts.

Climate Action Plymouth – Sustainable Power and Energy Group

10.37 Identified resources recoverable (tonnes of Uranium) of primary uranium amounts to 6.15 million tonnes. Australia is clearly a significant source for the near future. Reactor fuel is a mix of primary and secondary sources. Secondary sources include commercial and weapon stockpiles (about 15% of current reactor fuel), and from recycled plutonium and recycled and re-enriched uranium.

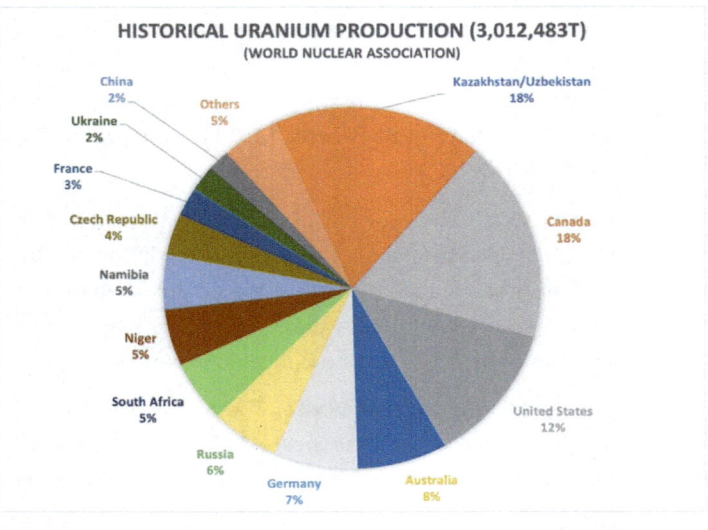

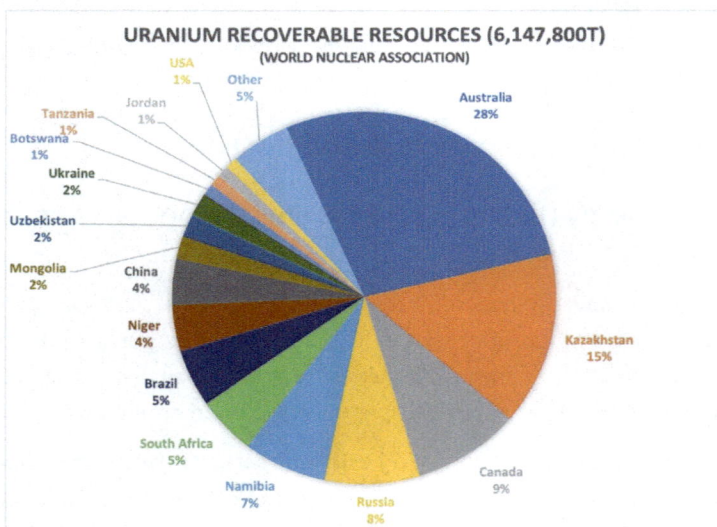

Historical and recoverable uranium resources by country

10.38 Safety remains a concern with the nuclear industry, something highlighted by the current situation in Ukraine. Mining of Uranium suffers the same safety issues as any other hard rock mining but in addition additional constraints are required to prevent inhalation of radionuclides. [279]

10.39 Nuclear power plant accidents involving a release of radioactive materials include [280].

- Major release – Chernobyl (Ukraine) [~ 45-200,000 radiation related deaths]
- Major release – Fukushima (Japan) [0-1 radiation related deaths].
- Significant release – Mayek (Russia) [~200 radiation related deaths]
- Limited release – Three Mile Island (USA) [no causal radiation related deaths]
- Limited release - Windscale (UK) [~ 33-100 radiation related deaths]

10.40 Each accident had a different cause. Fatalities and casualties vary enormously depending on the source.

10.41 Whilst **Friends of the Earth** does not support nuclear energy, it concludes on safety - "The health impacts of coal and gas are worse than from nuclear power, even with carbon capture and storage (CCS) in place. The health impacts for renewables, according to Tyndall, are broadly comparable to nuclear. But this assessment did not account for all health impacts resulting from nuclear accidents." [281] We will also oppose any life extensions to existing nuclear power plants if there are any significant safety concerns, or if they crowd out renewable power." "We do not oppose research into new potentially safer forms of nuclear power; but our current assessment is that we are extremely unlikely to need them in the future."

10.42 Nuclear fission power plants produce radioactive waste including plutonium and other actinides. Waste may be hazardous for thousands or hundreds of thousands of years. Long term disposal of waste is unresolved with no common agreement within the United Kingdom or worldwide on how it will be dealt with. [282]

10.43 Disposal of waste is a devolved matter. England and Wales favour deep geological whilst Scotland favours facilities near the surface. Northern Ireland is consulting along with the UK Government. The USA, France, Sweden, and Finland [283] also favour deep geological disposal or storage. No suitable sites have yet been identified.

10.44 Enriched uranium is used as fuel for the most common Light Water Reactors (LWR). The necessity for enrichment and the presence of plutonium in the spent fuel is the major proliferation concern. Building more nuclear reactors unavoidably increases nuclear proliferation risks because of its links to fuel enrichment and reprocessing. [284]

Fusion

10.45 **Fusion** is the opposite of fission. In fusion, energy is generated by bringing together two isotopes of hydrogen to make helium and in doing so to release heat which then produces steam to generate electricity. Whilst the recent announcement of successful fusion power is encouraging, fusion reactors have yet to produce more energy than they consume. [285] Tokamak Energy have been selected for Phase 1 AMR studies.

Innovation in Nuclear

10.46 In respect of new, innovative reactors, it's worth recalling Admiral Rickover's comments in 1953:

An academic reactor or reactor plant almost always has the following basic characteristics: It is simple. It is small. It is cheap. It is light. It can be built very quickly. It is very flexible in purpose ("omnibus reactor"). Very little development is required. It will use mostly "off-the-shelf" components. The reactor is in the study phase. It is not being built now.

On the other hand, a practical reactor plant can be distinguished by the following characteristics: It is being built now. It is behind schedule. It is requiring an immense amount of development on apparently trivial items. It is very expensive. It takes a long time to build because of the engineering development problems. It is large. It is heavy. It is complicated.

11. District Heating

District heating is currently configured mostly around maximising energy from Combined Heat and Power plant CHP. As such, whilst efficient, they are carbon emitting, often from wood or biomass.

District heating systems based on zero-carbon heat such as geothermal and/or ground source heat pumps are being introduced in Plymouth.

UK Government strategy and initiatives

11.1 The UK Government is supporting District Heating via its **Green Heat Network Fund** GHNF and its **Heat Networks Investment Project** HNIP.

PCC's CEAP initiatives

11.2 PCC is progressing the connection of the Civic Centre to the city centre low carbon heat network and assessing the feasibility of heat networks in Barne Barton and Derriford. [286]

11.3 PCC is assessing the feasibility of marine source heat pumps at various sites around Plymouth and testing the yield from ground source wells in Millbay to provide low carbon heat. [287]

District heating worldwide

11.4 District heating schemes are wide scale heating of large quantities of housing, industry, and business. Community heating schemes deliver to just one or two buildings.

11.5 District heating is used widely in Denmark where around 65% of all housing is served by district heating usually provided by not-for-profit organisations or consumer owned co-operatives. [288] Similarly high numbers exist in Sweden.

11.6 Fuels used for district heating are predominantly fossil fuel. [289] Finland is using a sand heat storage system to utilise excess wind and solar energy. [290]

11.7 There is over 4,444TWh of district heating worldwide. The largest uptake is in China and Russia, and in parts of Europe. The USA and UK have few district-heating schemes. [291]

FUEL FOR DISTRICT HEATING 2020 WORLDWIDE
HTTPS://WWW.IEA.ORG/REPORTS/DISTRICT-HEATING

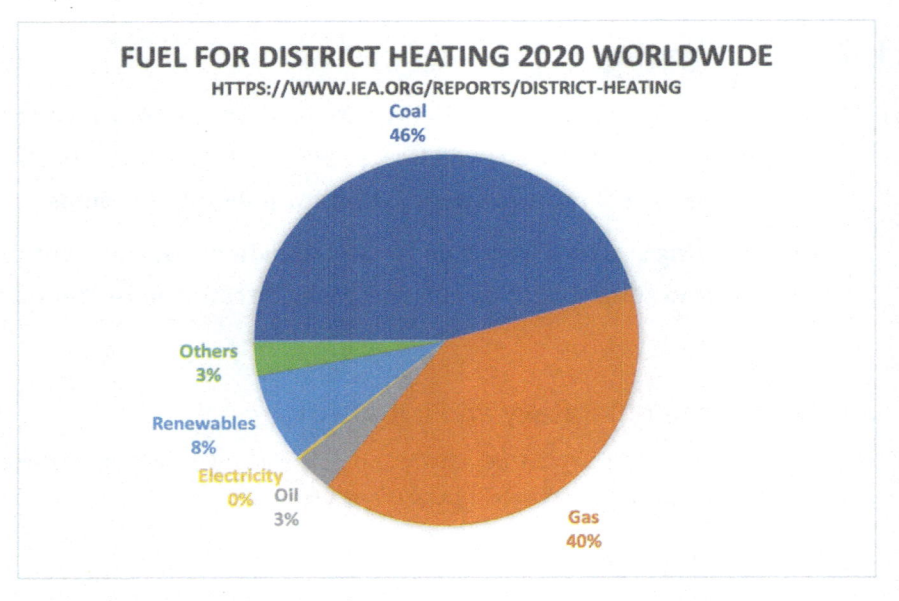

- Coal 46%
- Gas 40%
- Renewables 8%
- Oil 3%
- Others 3%
- Electricity 0%

District heating schemes today are largely powered by fossil fuel sources.

11.8 The **Green Heat Network Fund** [292] is a 3-year £288m capital grant fund that will support the commercialisation and construction of new low and zero carbon (LZC) heat networks (including the supply of cooling), and the retrofitting and expansion of existing heat networks. It is open only to organisations in the public, private, and third sector, and not to individuals, households, and sole traders. The scheme will close in 2025.

11.9 The **Heat Networks Investment Project** (HNIP) [293] is a government funding programme to increase the number of heat networks being built, deliver carbon savings, and create the conditions necessary for a sustainable heat network market to develop. £320m of funds are available. The scheme is open to all. Most grants made by HNIP are for waste incineration, biomass, or gas Combined Heat and Power CHP schemes. [294] Some schemes have an ambition to move to zero carbon fuels.

11.10 In the **UK, district heating** accounts for just 2% of all domestic heating. This amounts to 12,000GWh with 6,500 GWH going to domestic users and 5,500 GWh going to non-domestic users. [295] There are over 17,000 district heat networks in the UK with over 500,000 connections to them.

11.11 Initially in the UK, district heating was local and provided by burning fossil fuels, primarily gas and wood, or from incinerators. Efficient but still carbon emitting, Combined Heat and Power CHP schemes still make up most district heating schemes.

11.12 The most recent schemes are much larger and usually provide heat from renewable or low carbon heat sources – nuclear (geothermal nuclear decay and nuclear waste), heat pumps (ground, marine and air), and landfill heat. With the adoption of heat pumps, any low-grade heat source can be utilised.

11.13 **Plymouth is introducing pilot district heating schemes** in two clusters – Civic and Millbay. [296 297]

11.14 The **Civic Scheme** uses 71kWp of solar PV delivering 66.5 MWh/year and saving 5t CO_2 pa and a second 200kWp ASHP delivering 354MWh per annum and saving 88t CO_2 pa.

11.15 The **Millbay Scheme** uses 67kWp Solar PV, delivering 88MWh pa and saving 24t CO_2pa. A second scheme uses 200kWp ASHP delivering 795MWh pa and saving 138t CO_2pa.

11.16 Future nation-wide schemes are expected to include:

- **carbon sources**: - large biomass, municipal waste, and waste heat recovered from a combined heat and power plant.
- **non-carbon sources**: waste heat from industrial activities, geothermal energy, heat latent in the environment, such as in rivers and lakes, combined with large scale heat pumps, large scale solar thermal heat, and extraction from the sewage networks.

Environmental impact of district heating

11.17 The source of heat for a district heating scheme will have an environmental impact depending on its fuel. As well as the fuel impact, the other subsystems include power plant, main grid, auxiliary components of the main grid, building works, service pipes, and dwelling infrastructure. [298]

12. Hydrogen

Hydrogen should only be produced from excess zero-emission renewables. There is little likelihood that large amounts of excess electricity for hydrogen production will be available in the near term.

Given the production process, using electricity to produce the gas and then the gas to produce electricity, costs per kWh will be high.

Hydrogen should not be used for combustion, including in domestic boilers, as toxic emissions from domestic hydrogen boilers will be higher than with natural gas.

UK Government strategy and initiatives

12.1 The UK Government in its **British Energy Security Strategy (BESS)** projects up to 10GW of hydrogen by 2030, half green and half blue. Up to 20% of hydrogen will be blended into the gas grid.

12.2 The **Heat and Buildings Strategy** (H&BS) aims to establish the feasibility of hydrogen, undertake large scale trials, enable hydrogen and natural gas blending, and consult on hydrogen ready boilers.

Local developments

12.3 As part of Plymouth's proposed 'freeport', a 10MW green hydrogen hub project at Langage Energy Park has been proposed by Carlton Power. Carlton Power developed the Langage gas-fired power station and solar farm. Planning consent has been given by South Hams District Council. It is expected to provide local companies with easy access to hydrogen fuel. The Government's Hydrogen Investment Package provides the funding. Hydrogen at scale is expected within the next two to three years. [299 300]

Hydrogen for home heating

12.4 "Compared to other alternatives such as heat pumps, solar thermal, and district heating, hydrogen use for domestic heating is less economic, less efficient, more resource intensive, and associated with larger environmental impacts." - Dr. Jan Rosenow, Director of European programs at the Regulatory Assistance Project (RAP) [301]

12.5 "In the UK, heating homes with green hydrogen would use approximately six times more renewable electricity than heat pumps," - David Cebon of the Hydrogen Science Coalition and Professor of Mechanical Engineering in Cambridge University. [302]

12.6 Cornwall Insight consultancy warns "that replacing gas with hydrogen for home heating would result in huge energy bill rises for households over the next 30 years, with one estimate putting the increase at between 70-90%." [303]

Hydrogen in a fuel cell

12.7 Hydrogen has the potential to reduce climate change if it is both produced and used in a zero-carbon manner, such as a fuel cell. A fuel cell produces electricity when fuel is provided. Fuel cells have an anode and a cathode. Hydrogen is fed to the anode which, using a catalyst, separates the hydrogen atom into protons and electrons. The electrons then flow to an external circuit to produce electricity. Protons flow to the cathode where they recombine with oxygen to produce water. [304]

12.8 **Fuel cells** have a potential application in transport, particularly in larger vehicles and marine vessels, but their use is not yet widespread. Stationary emergency power is another application, although the price of batteries has undercut this market.

12.9 Such is the inefficiency and costs of producing hydrogen that direct battery electric should be used whenever feasible. Where it is not feasible, then hydrogen can be used as a transport fuel, for heavy goods vehicles, for longer range marine vessels and for larger, longer-range aircraft but it may no longer be carbon neutral. If hydrogen is produced using fossil fuels and then burnt it is carbon positive.

12.10 Hydrogen can also be used as an energy store, by utilising excess renewable energy and storing it as hydrogen before using it in a fuel cell to produce electricity. However, this is much less efficient than most other energy storage systems although it does have the benefit of longer-term storage.

Hydrogen production

12.11 **Green hydrogen**, when produced by electrolysis, is dependent upon a readily available source of renewable energy or will depend upon the carbon intensity of grid electricity, similarly to EVs. However, the process thereafter significantly worsens the carbon intensity – transport of the fuel and the inefficiency of fuel cells when compared to battery electric becomes the problem.

12.12 **Pink hydrogen** is produced from electrolysis powered by nuclear power.

12.13 **Grey hydrogen** is hydrogen produced from methane by steam reformation.

12.14 **Blue hydrogen** is grey hydrogen but including the unproven benefits of Carbon Capture and Storage CCS.

12.15 **Both grey and blue hydrogen significantly worsen the carbon credentials of hydrogen.** Blue or grey hydrogen are not sustainable and insufficient carbon free electricity exists to produce green hydrogen in the required quantities.

Hydrogen for heating

12.16 Using hydrogen in a domestic boiler requires combustion where, because hydrogen is not a hydrocarbon, there is no carbon dioxide emitted. Hydrogen heating will likely be as expensive as direct electric heating.

12.17 There will be toxic emissions from the combustion process. [305] Burning hydrogen in pure oxygen will produce just water. Burning hydrogen in air (21% oxygen and 78% nitrogen by volume) will produce nitrogen dioxide NOx and, because there is no carbon available, NOx will be higher than for fossil fuels.

12.18 Hydrogen will require a new infrastructure. This is likely to be expensive.

Hydrogen heating efficiency.

12.19 Given that renewable energy and nuclear are the core energy technologies of the future, it is helpful to look at the efficiency of various ways of heating homes with the electricity produced. Options include heat pump (180-460%), direct heating (92%), and hydrogen (57%).

Heat Pump
Renewable Energy – Transmission (92%) – Heat Pump (200-500%) – **Energy Efficiency 180-460%**

Direct Electric Heating
Renewable Energy – Transmission (92%) – Direct Electric (100%) -   **Energy Efficiency 92%**

Hydrogen Boiler
Renewable Energy – Electrolyser (74%) – Gas Grid (97%) - Boiler (80%) - **Energy Efficiency 57%**

13. Conclusions

13.1 Climate change is a result of increasing greenhouse gases in the atmosphere. Greenhouse gases have been increasing since the beginning of the industrial revolution but have gathered pace since the middle of the last century - they continue to increase even now and must be curtailed if global temperatures are not to rise to levels where the collapse of ecological systems leads to mass extinction of flora and fauna and societal breakdown due to water and food shortages causing high levels of migration and the inevitable conflict that will bring. Eventually, rising temperatures will render the planet uninhabitable for life as we know it.

13.2 The most significant contributor to greenhouse gases is carbon dioxide which is the direct result of the combustion of fossil fuels and biomass to produce heat, power, and energy. In Great Britain, methane, the major constituent of natural gas, is the most significant fossil fuel being combusted (or burnt) to produce heat, power, and energy. As well as carbon dioxide exhaust emissions, methane escapes during drilling and during transport, as well as during the initial stages of combustion. Clearly, we must 'stop burning things' and we must stop using natural gas as the fuel that is burnt.

13.3 The combustion of any fuel in air, rather than in pure oxygen, will lead to toxic exhaust emissions including nitrogen oxides and particulates. This applies to both carbon fuels (fossil and biomass) and non-carbon fuels, such as hydrogen. Combustion of fuel is not the answer either for the long term or as a transition.

13.4 Since abundant natural gas became cheaply available in almost every home, low gas prices have led to its almost universal adoption in Great Britain for home heating and hot water with little regard for improving efficiency by improving home insulation and building standards. Recent high gas prices have focussed homeowners' attention on lowering energy demand and on maximising efficiency from poorly installed condensing (and particularly combi) boilers. High gas prices may well be a driver to lower greenhouse and toxic emissions. Improving insulation and correctly setting up a condensing boiler is a low cost and very effective means of reducing personal carbon emissions with the added benefit of reducing household bills.

13.5 It was only just over a decade ago when electricity generation was dominated by coal together with a significant nuclear input. Coal has now almost entirely been replaced by natural gas and renewables whilst the building of new nuclear has been a casualty of the energy free market. As a result, the carbon intensity of the grid electricity is now considerably lower now than a decade ago, is lower than energy produced from a domestic boiler, and substantially lower than energy derived from Liquified Natural Gas, yet electricity still counts as a high carbon energy source, meaning it attracts extra taxation and levies to its retail price whilst failing to qualify for many improvement grants. The government must follow through on its intention to bring the unit gas price more in line with low carbon electricity. Part of this solution is to de-couple the renewable electricity price from the price driven by gas generated electricity. Aligning electricity and gas prices is of the utmost importance if the transition to electric heat pumps is to gain the momentum necessary to move home heating and hot water from high emission gas to low carbon electricity.

13.6 Great Britain's total energy demands are currently met by around 87% combusted fuels (gas, bio, petroleum) and just 13% from combustion-free sources. Electricity delivers around 25% of our annual total energy demand, almost half of which is derived from the combustion of fossil and biofuels. By 2050, if net-zero is to be reached, the energy required for heating, hot water, electricity generation, and transportation will need to be delivered through electricity produced by combustion-free power generation. Combustion of fuels of any kind must have ended before then.

13.7 To be able to meet an all-electric future powered by combustion-free sources, it will be of the utmost importance to simultaneously reduce energy demand by around 30% otherwise achieving this target by 2050 will likely be beyond reach. Plymouth should require all new buildings to have a high standard of insulation and ventilation and encourage all existing dwellings to have as high a standard as reasonably practicable. Local Authority buildings should have improvement plans to achieve as high a carbon efficiency as is achievable, something that is already underway in the city.

13.8 Power for that all-electric future from combustion-free sources (renewables together with nuclear) must grow from just 10% of current total energy to near 100% - a five-fold increase in electricity generated but a ten to fifteen-fold increase in installed capacity when considering renewable derating for intermittency. The most optimistic scenario from the National grid is a four-fold increase in combustion-free renewables energy generation so there is still quite some way to go. Plymouth should require that all new dwellings are fitted with battery backed rooftop solar and that locations for suitable wind and solar farms are fast-tracked wherever possible.

13.9 Of great importance to the transition to an all-electric low-carbon future is the infrastructure necessary to connect the new renewable sites to the grid. This requires a substantial investment of about the same level as that of the last thirty years. The government has taken the steps to part nationalise the grid to ensure the necessary funding is available but pressure to complete rather than defer this investment will be essential. There are substantial jobs and growth potential here.

13.10 Wind, both onshore and offshore, and large-scale solar are the cheapest forms of energy available. They are low risk and offer a quick return on capital employed. In Scotland, onshore wind is significant with a growing amount of offshore in England. New onshore wind in England has been curtailed by government legislation that makes new onshore wind installations all but impossible to implement. The government must change their position on this and must not impose the same constraints on large scale solar: it is impossible to meet net zero with offshore wind alone. Hydro, tidal stream, wave and geothermal energy all offer forms of combustion-free renewable energy but, whilst it is important to continue to seed fund these technologies, they cannot be relied upon to scale up sufficiently quickly enough to make an impact in the short timescales necessary. **Technologies that can be scaled up today are the ones that will be able to make the impact necessary in the timescale available.**

13.11 Improving the dependency of renewables and addressing their intermittency can be in large part achieved through grid-scale energy storage of which battery storage is one technology that can be scaled up at relatively low cost. Other storage technologies are also being developed.

13.12 Nuclear fuelled steam power generation was a British success, but the freeing up of the energy market ensured no new reactors have been built and become operational for some decades. Nuclear is a high risk, high capital cost investment technology with a very long payback time. A Hinkley C model of investment has not yet delivered any power and when it does so the energy unit cost will be high. Sizewell C, if it is ever built, will be funded as a Regulated Asset Base making the electricity produced a little cheaper but at a higher risk of cost and timescale overrun to the taxpayer. The government's aim of a further eight nuclear power generating stations seems a very long way off.

13.13 Small/Advanced Modular Reactors may well be part of the energy solution. Factory built, they may be quicker to generate power although as no such plant are yet approved, their contribution to net zero in the timescale necessary must be in doubt.

13.14 Hydrogen is often promoted as a clean fuel, but its production is energy intensive and hence carbon intensive if it is not done by zero carbon renewables. Using hydrogen in a fuel cell is emission free but burning or combusting hydrogen produces toxic emissions, something which will further exacerbate our air quality issues. Using hydrogen in today's domestic gas boilers in now firmly out of favour. Hydrogen is also almost certain to be more expensive than both gas and electricity and will do little to address energy poverty.

13.15 Net zero is achievable but it requires a programme of energy efficient insulation and ventilation, an affordable change from gas heating to electric heat pumps, and a significant switch from combustion fuels to electricity generated from wind and solar.

13.16 Personal contributions include improving insulation, electrifying heating and cooking, and upgrading to more efficient appliances. Using off-peak electricity, avoiding the peak 4-7pm slot, will ensure the lowest carbon electricity is used.

14. About the Author and Climate Action Plymouth

14.1 David Bricknell CEng FRINA BSc(Hons) is an engineer with some 50 years of experience in the subject. During his career he has been responsible for R&D and systems design in Power and Propulsion which has provided the knowledge and expertise to write two books for Climate Action Plymouth: *Combustion-Free Mobility* and this book on *Combustion-Free Heat, Power, and Energy for Plymouth*. He has also written many books on Electric Vehicle Technologies. David is passionate about environmental issues and the alternative technical solutions necessary to avoid or mitigate the climate issues associated with the combustion of fossil and other fuels. He lives in Plymouth and is married with two children and three grandchildren.

14.2 Climate Action Plymouth is a grass-roots movement affiliated to Friends of the Earth. We exist to bring together people and organisations to promote positive change in this time of global climate crisis. We focus on actions to reduce greenhouse gas emissions locally and nationally to achieve true net zero by 2030.

15. References

[1] Covering Climate Now https://coveringclimatenow.org

[2] https://www.theguardian.com/uk-news/2022/jul/13/sir-patrick-vallance-gives-emergency-climate-briefing-to-uk-mps?CMP=Share_iOSApp_Other

[3] World Meteorological Organization COP27 https://www.theguardian.com/environment/2022/nov/06/climate-crisis-past-eight-years-were-the-eight-hottest-ever-says-un

[4] Sir Patrick Vallance gives emergency briefing to UK MPs – The Guardian 13th July 2022
https://www.gov.uk/government/publications/british-energy-security-strategy/british-energy-security-strategy

[5] Government Policy Paper Heat and Buildings Strategy https://www.gov.uk/government/publications/heat-and-buildings-strategy

[6] Plymouth City Council – Climate Emergency Action Plan 2022
https://www.plymouth.gov.uk/environmentandpollution/climateemergency/climateemergencyactionplan/climateemergencyactionplan2022

[7] Intergovernmental Panel on Climate Change (IPCC) Sixth Assessment Report https://www.ipcc.ch/report/ar6/wg3/

[8] Government Policy Paper – British Energy Security Strategy (BESS) https://www.gov.uk/government/publications/british-energy-security-strategy

[9] 2018 Greenhouse Gas Emissions, provisional figures - Statistical Release: National Statistics
https://assets.publishing.service.gov.uk/government/uploads/system/uploads/attachment_data/file/790626/2018-provisional-emissions-statistics-report.pdf

[10] Scientists have just told us how to solve the climate crisis – will the world listen – The Guardian 6th April 2022
https://www.theguardian.com/commentisfree/2022/apr/06/scientists-climate-crisis-ipcc-report

[11] American Meteorological Society State of the Climate report 20212 https://www.ametsoc.org/ams/index.cfm/publications/bulletin-of-the-american-meteorological-society-bams/state-of-the-climate/

[12] Plymouth City Council - Climate Emergency Public Consultation
https://www.plymouth.gov.uk/environmentandpollution/climatechangeandenergy/climateemergencypublicconsultation

[13] Late Ordovician mass extinction caused by volcanism, warming and anoxia. David P.G. Bond and Stephen E. Grasby - Geology (2020) 48 (8): 777-781
https://pubs.geoscienceworld.org/gsa/geology/article/48/8/777/586486/Late-Ordovician-mass-extinction-caused-by

[14] A long-term association between global temperature and biodiversity, origination and extinction in the fossil record. Peter J Mayhew, GarethB Jenkins, and Timothy G Benton. Proc Biol Sci 2008 Jan 7; 275 (1630): 47-53..
https://www.ncbi.nlm.nih.gov/pmc/articles/PMC2562410/

[15] Anomalous K-Pg-aged seafloor attributed to impact-induced mid-ocean ridge magmatism. Joseph S. Byrnes and Leif Karlstrom – Science advances Vol 4 Issue 2 7 February 2018 https://www.science.org/doi/10.1126/sciadv.aao2994

[16] Discovery of the Greenhouse Effect – Richard Lemmons, Climate Policy Watcher August 2022 https://www.climate-policy-watcher.org/greenhouse-gases-2/discovery-of-the-greenhouse-effect.html

[17] UC San Diego, Scripps Institution of Oceanography – The Keeling Curve https://keelingcurve.ucsd.edu

[18] United Nations IPCC report 'Code red' for human driven global heating warns UN chief – 9th August 2021.
https://news.un.org/en/story/2021/08/1097362

[19] IPCC Sixth Assessment Report 4th April 2022 https://www.ipcc.ch/assessment-report/ar6/

[20] IPCC Climate Change 2022 – Mitigation of Climate Change, Summary for Policymakers
https://report.ipcc.ch/ar6wg3/pdf/IPCC_AR6_WGIII_SummaryForPolicymakers.pdf

[21] The Arctic is now expected to be ice-free by 2040 – World Economic Forum, 17th May 2017
https://www.weforum.org/agenda/2017/05/the-arctic-could-be-ice-free-by-2040/

[22] Global Heating: best and worst case scenarios less likely than thought – The Guardian
https://www.theguardian.com/environment/2020/jul/22/global-heating-study-narrows-range-of-probable-temperature-rises

[23] Atmospheric implications of increased Hydrogen use, Nicola Warwick, Paul Griffiths, James Keeble, Alexander Archibald, John Pyle –

Climate Action Plymouth – Sustainable Power and Energy Group

University of Cambridge and NCAS and Keith Shine - University of Readings April 2022
https://assets.publishing.service.gov.uk/government/uploads/system/uploads/attachment_data/file/1067144/atmospheric-implications-of-increased-hydrogen-use.pdf

[24] Direct Emissions from Stationary Combustion Sources – Greenhouse Gas Inventory Guidance – USA Environmental Protection Agency
https://www.epa.gov/sites/default/files/2020-12/documents/stationaryemissions.pdf

[25] Nitrous Oxide Greenhouse Gas Inventory summary Factsheet,
https://assets.publishing.service.gov.uk/government/uploads/system/uploads/attachment_data/file/48423/5557-nitrous-oxide-factsheet.pdf

[26] Effect on NOx Emissions of Hydrogen Addition to Natural Gas in Industrial Boilers and Heaters, Peter Gogolek, Research Scientist, Industrial Innovation Group, March 2021 – Department of Energy and Climate Change – Natural Resources Canada
http://publicsde.regie-energie.qc.ca/projets/593/DocPrj/R-4165-2021-B-0019-DemAmend-Autre-2021_08_13.pdf

[27] Atmospheric implications of increased Hydrogen use, Nicola Warwick, Paul Griffiths, James Keeble, Alexander Archibald, John Pyle – University of Cambridge and NCAS and Keith Shine - University of Readings April 2022
https://assets.publishing.service.gov.uk/government/uploads/system/uploads/attachment_data/file/1067144/atmospheric-implications-of-increased-hydrogen-use.pdf

[28] Analysis: Why the UK's CO2 emissions have fallen 38% since 1990 – Carbon Brief, Clear on Climate
https://www.carbonbrief.org/analysis-why-the-uks-co2-emissions-have-fallen-38-since-1990

[29] The decoupling of economic growth from carbon emissions: UK evidence, Census 2021
https://www.ons.gov.uk/economy/nationalaccounts/uksectoraccounts/compendium/economicreview/october2019/thedecouplingofeconomicgrowthfromcarbonemissionsukevidence

[30] 2020 UK greenhouse gas emissions, provisional figures – Department for Business, Energy & Industrial Strategy BEIS, 25th March 2021
https://assets.publishing.service.gov.uk/government/uploads/system/uploads/attachment_data/file/972583/2020_Provisional_emissions_statistics_report.pdf

[31] Plymouth worse than London for air pollution – World Health Organisation - Plymouth Live
https://www.plymouthherald.co.uk/news/plymouth-worse-london-air-pollution-1523721

- [32] WHO Global Urban Ambient Air Pollution Database (update 2016) Ambient (outdoor) air pollution database, by country and city
xlsx, 1.31Mb

[33] Airborne particulate matter and their health effects – Encyclopedia of the Environment https://www.encyclopedie-environnement.org/en/health/airborne-particulate-health-effects/

[34] National Air Quality Testing Services https://www.naqts.com/whats-in-your-air/particles/

[35] Reports and statements from the Committee on the Medical Effects of Air Pollutants (COMEAP).
https://www.gov.uk/government/collections/comeap-reports

[36] WHO limits for particulate matter will be enshrined in UK law, pledges Gove – air quality news
(https://airqualitynews.com/2019/07/16/who-limits-for-particulate-matter-will-be-enshrined-in-uk-law-pledges-gove/)

[37] Plymouth's polluted air is dangerous to breathe, World Health Organisation report claims – Plymouth Live
(https://www.plymouthherald.co.uk/news/health/plymouths-polluted-air-dangerous-breathe-704029)

[38] Air quality in Plymouth - IQAir https://www.iqair.com/uk/england/plymouth

[39] Power Compare https://powercompare.co.uk/uk-energy-industry/

[40] Office of Gas and Electricity Markets OFGEM https://www.ofgem.gov.uk/energy-advice-households/costs-your-energy-bill

[41] Government Policy Paper – Heat and Buildings Strategy 19th October 2021 https://www.gov.uk/government/publications/heat-and-buildings-strategy

[42] OFGEM Energy Company Obligation (ECO) https://www.ofgem.gov.uk/environmental-and-social-schemes/energy-company-obligation-eco

[43] OFGEM Energy Price Cap Explained 1st October 2022. https://www.ofgem.gov.uk/information-consumers/energy-advice-households/check-if-energy-price-cap-affects-you

Combustion-Free Heat, Power, and Energy

[44] OFGEM Insights paper on households with electric and other non-gas heating. 1th December 2015
https://www.ofgem.gov.uk/publications/insights-paper-households-electric-and-other-non-gas-heating

[45] IEA Contracts For Difference 25th October 2019 https://www.iea.org/policies/5731-contract-for-difference-cfd

[46] Government Consultation – Review of electricity market arrangements 18th July 2022
https://www.gov.uk/government/consultations/review-of-electricity-market-arrangements

[47] Energy Voice PM's Plan needs to do more to decouple gas from electricity prices 13/09/2022
https://www.energyvoice.com/oilandgas/443692/uk-decouple-gas-electricity/

[48] Government Press Release – Government hits accelerator on low-cost renewable power 9th February 2022
https://www.gov.uk/government/news/government-hits-accelerator-on-low-cost-renewable-power

[49] Government Guidance – Boiler Upgrade Scheme 23rd May 2022 https://www.gov.uk/guidance/check-if-you-may-be-eligible-for-the-boiler-upgrade-scheme-from-april-2022

[50] Dept for Business Energy & Industrial Strategy Green Heat Network Fund Version 3
https://assets.publishing.service.gov.uk/government/uploads/system/uploads/attachment_data/file/1076541/ghnf-r1-scheme-overview.pdf

[51] Government Notice Apply for the Social Housing Decarbonisation Fund: Wave 2.1 7th October 2022
https://www.gov.uk/government/publications/social-housing-decarbonisation-fund-wave-2

[52] Yes Energy Solutions – What is the Home Upgrade Grant https://www.yesenergysolutions.co.uk/advice/home-upgrade-grant-(hug)/what-is-the-home-upgrade-grant/

[53] RIBA The Future Homes Standard explained https://www.architecture.com/knowledge-and-resources/knowledge-landing-page/the-future-homes-standard-explained

[54] Engineering and Technology – Government announces £1.5bn insulation scheme for low-income households September 29 2022
https://eandt.theiet.org/content/articles/2022/09/government-announces-15bn-insulation-scheme-for-low-income-households/

[55] Government Consultation Delivering a smart and secure electricity system: the interoperability and cyber security of energy smart appliances and remote load control 6 July 2022 https://www.gov.uk/government/consultations/delivering-a-smart-and-secure-electricity-system-the-interoperability-and-cyber-security-of-energy-smart-appliances-and-remote-load-control

[56] Plymouth Energy Community - Community led housing at Kings Tamerton. https://plymouthenergycommunity.com/our-work/kings-tamerton?r=1808

[57] Eurostat Statistics explained – Energy consumption in households June 2022 https://ec.europa.eu/eurostat/statistics-explained/index.php?title=Energy_consumption_in_households

[58] Energy Saving Trust 2022 – Top five energy consuming home appliances https://energysavingtrust.org.uk/top-five-energy-consuming-home-appliances/

[59] Energy Consumption in the UK (ECUK) 1970 to 2019 Dept for Business, Energy & Industrial Strategy – National Statistics.

[60] RIBA – Homes for Heroes 2022 https://www.architecture.com/knowledge-and-resources/resources-landing-page/homes-for-heroes

[61] Nuaire Mechanical Ventilation Heat Recovery System MVHR - Llyr Davies October 2014 https://www.nuaire.co.uk/blog/mvhr-most-frequently-asked-questions#flex-4c4a4fb9-d1aa-41bd-b150-3fb6fde605dd

[62] Passivhaus Trust – What is Passivhaus https://www.passivhaustrust.org.uk/what_is_passivhaus.php

[63] Passivhous Trust Primrose Park claim's UK's largest residential Passivhaus title.
https://www.passivhaustrust.org.uk/news/detail/?nId=786

[64] Energie Sprong – This Dutch construction innovation shows it's possible to quickly retrofit every building
https://energiesprong.org/this-dutch-construction-innovation-shows-its-possible-to-quickly-retrofit-every-building/

[65] Energir Sprong – Desirable, warm, affordable homes for life. https://www.energiesprong.uk

[66] Greenleaf Road Retrofit Factsheet.pdf

[67] Homes for Heroes 2022 RIBA Architecture

[68] Plymouth City Council – Housing Plan 2012-2017
https://democracy.plymouth.gov.uk/documents/s37045/Housing%20Plan%20Final.pdf

[69] Plymouth City Council – Housing Plan 2012-2017
https://democracy.plymouth.gov.uk/documents/s37045/Housing%20Plan%20Final.pdf

[70] Plymouth City Council – Low Carbon Team.

[71] Interreg North West Europe HeatNet NEW Plymouth Pilot Case Study https://guidetodistrictheating.eu/plymouth/

[72] Which – Which boilers are most energy efficient 25 Feb 2022 https://www.which.co.uk/reviews/boilers/article/boiler-energy-efficiency-aCgnH9h8JJP9

[73] The Heating Hub – Boiler efficiency calculator – Most Efficient boilers – energy saving tips. https://www.theheatinghub.co.uk/boiler-efficiency-guide-and-energy-saving-tips

[74] The Heating Hub – Why our condensing boilers do not condense.05/03/2020 https://www.theheatinghub.co.uk/why-our-condensing-boilers-do-not-condense

[75] Energy Saving Trust, Gastec at CRE, EA Technology, AECOM, "In-situ monitoring of efficiencies of condensing boilers and use of secondary heating trial - final report," June 2009.

[76] The Heating Hub – Save up to 12% in gas use, by turning down the 'flow' temperature. 10/10/2021
https://www.theheatinghub.co.uk/articles/turn-down-the-boiler-flow-temperature

[77] Electrification of heat BEAMA October 2021

[78] Energy Saving Trust – Ground Source Heat Pumps https://energysavingtrust.org.uk/advice/ground-to-water-heat-pumps/

[79] Energy Saving Trust – In-depth guide to heat pumps https://energysavingtrust.org.uk/advice/in-depth-guide-to-heat-pumps/

[80] The Eco Experts – High Temperature Air Source Heat Pumps https://www.theecoexperts.co.uk/blog/2022/01/28/high-temperature-air-source-heat-pumps

[81] Homebuilding and Renovating High Temperature Heat Pumps: Are they a Eureka product for homeowners? https://www.homebuilding.co.uk/advice/high-temperature-heat-pumps

[82] Vattenfall launches high-temperature heat pumps that can directly replace gas boilers. https://www.current-news.co.uk/news/vattenfall-launches-high-temperature-heat-pumps-that-can-directly-replace-gas-boilers

[83] Electrification of Heat Demonstration Project – Dept for Business, Energy & Industrial Strategy.

[84] Vattenfall's new heat pump provides homeowners an alternative to gas 18 November 2021 https://group.vattenfall.com/press-and-media/newsroom/2021/vattenfalls-new-heat-pump-provides-homeowners-an-alternative-to-gas

[85] UN Climate body accepts GWP of R1234yf refrigerant is 'lower than CO_2'. https://www.racplus.com/news/un-climate-body-accepts-gwp-of-r1234yf-refrigerant-is-lower-than-co2-10-02-2014/

[86] RAC Daimler will phase in CO2 from next year (2015) https://www.racplus.com/news/daimler-will-phase-in-co2-from-next-year-picture-special-15-12-2015/

[87] New generation refrigerants for domestic heat pumps in Sweden - Martina Longhini

[88] Tepeo Zero Emission Boiler (ZEB) https://tepeo.com/thezeb

[89] The impact of interconnectors on decarbonisation Department of Business, Energy and Industrial Strategy October 2020 BEIS Research Paper number 2020/056
https://assets.publishing.service.gov.uk/government/uploads/system/uploads/attachment_data/file/943239/impact-of-interconnectors-on-decarbonisation.pdf

[90] MyGridGB http://www.mygridgb.co.uk/dashboard/

[91] Dept of BEIS – Digest of UK Energy Statistics – Annual data for UK, 2020.
https://assets.publishing.service.gov.uk/government/uploads/system/uploads/attachment_data/file/1060151/DUKES_2021_Chapters_1_to_7.pdf

[92] nationalgridESO Net Zero Future Energy Scenarios 2022 https://www.nationalgrideso.com/future-energy/future-energy-scenarios/fes-2021/scenarios-net-zero

[93] National Gris to be partly nationalised to help reach net zero targets 6 April 2022 The Guardian
https://www.theguardian.com/business/2022/apr/06/national-grid-to-be-partially-nationalised-to-help-reach-net-zero-targets

Combustion-Free Heat, Power, and Energy

[94] Gridwatch http://www.gridwatch.templar.co.uk

[95] Energy Saving Trust 26 August 2020 Time of use tariffs: all you need to know https://energysavingtrust.org.uk/time-use-tariffs-all-you-need-know/

[96] Carbon Intensity API National Grid ESO, Environmental Defense Fund, University of Oxford Department of Computer Science and WWF https://www.carbonintensity.org.uk

[97] Government National Statistics Digest of UK Energy Statistics (DUKES): natural gas https://www.gov.uk/government/statistics/natural-gas-chapter-4-digest-of-united-kingdom-energy-statistics-dukes

[98] Houses of Parliament May 2016 Number 523 Carbon Footprint of Heat Generation https://researchbriefings.files.parliament.uk/documents/POST-PN-0523/POST-PN-0523.pdf

[99] OFGEM Ofgen reveals landmark five-year programme to deliver reliable, sustainable energy at the lowest cost to consumers 29 June 2022 https://www.ofgem.gov.uk/publications/ofgem-reveals-landmark-five-year-programme-deliver-reliable-sustainable-energy-lowest-cost-consumers

[100] Department for Business, Energy and Industrial Strategy - Electricity Generation Costs 2020 https://assets.publishing.service.gov.uk/government/uploads/system/uploads/attachment_data/file/911817/electricity-generation-cost-report-2020.pdf

[101] The Crown Estate 03 July 2018 Electricity interconnectors https://www.thecrownestate.co.uk/en-gb/media-and-insights/stories/2018-electricity-interconnectors/

[102] Energy Voice – Xlinks launches Morocco-UK renewable energy plan. 26/09/2021 https://xlinks.co/morocco-uk-power-project/ https://www.energyvoice.com/renewables-energy-transition/solar/352090/xlinks-morocco-uk-interconnectors/

[103] Government Press Release 23 February 2022 – Government boost for new renewable energy storage technologies https://www.gov.uk/government/news/government-boost-for-new-renewable-energy-storage-technologies

[104] https://www.gov.uk/government/publications/longer-duration-energy-storage-demonstration-programme-successful-projects/longer-duration-energy-storage-demonstration-programme-stream-2-phase-1-details-of-successful-projects

[105] https://www.rheenergise.com/press-release---beis

[106] BBC The massive green power projects stuck in limbo 24 June 2021. https://www.bbc.co.uk/news/uk-scotland-highlands-islands-57510870

[107] Energy Storage for net zero by 2050 National Grid https://www.energy-storage.news/uk-needs-at-least-50gw-of-energy-storage-for-net-zero-by-2050-national-grid-eso-says/

[108] Intergen gains consent to build one of the world's largest battery projects in Essex 30 Nov 2020 https://www.intergen.com/news-insights/categories/news/intergen-gains-consent-to-build-one-of-the-world-s-largest-battery-projects-in-essex/

[109] Top 5 largest battery energy storage systems worldwide https://www.saurenergy.com/solar-energy-news/the-top-5-largest-battery-energy-storage-systems-worldwide

[110] https://www.energy-storage.news/uk-needs-at-least-50gw-of-energy-storage-for-net-zero-by-2050-national-grid-eso-says/

[111] Government Press Release – Battery storage boost to power greener electricity grid. https://www.gov.uk/government/news/battery-storage-boost-to-power-greener-electricity-grid

[112] Domestic battery energy storage systems https://www.gov.uk/government/publications/domestic-battery-energy-storage-systems

[113] Office for National Statistics ONS – Families and households in the UK: 2020 https://www.ons.gov.uk/peoplepopulationandcommunity/birthsdeathsandmarriages/families/bulletins/familiesandhouseholds/2020

[114] nationalgridESO EVs and electricity 2022 https://www.nationalgrideso.com/uk/electricity-transmission/future-energy/net-zero-explained/electric-vehicles/evs-electricity

[115] OFGEM – Case study (UK): electric vehicle-to-grid (V2G) charging 6 July 2021 https://www.ofgem.gov.uk/publications/case-study-uk-electric-vehicle-grid-v2g-charging

[116] Energy Superhub Oxford – UK's largest flow battery energised at Energy Superhub Oxford 20 December 2021 https://energysuperhuboxford.org/uks-largest-flow-battery-energised-at-energy-superhub-oxford/

Climate Action Plymouth – Sustainable Power and Energy Group

[117] USA Dept of Energy - Using hot sand to store energy 31 August 2021
https://cleantechnica.com/2021/08/31/using-hot-sand-to-store-energy/

[118] Science Direct – Compressed Air Energy Storage, Paul Breeze, in Power System Energy Storage Technologies, 2018
https://www.sciencedirect.com/topics/engineering/compressed-air-energy-storage

[119] UN Climate Technology Centre & Network – Compressed Air Energy Storage (CAES)
https://www.ctc-n.org/technologies/compressed-air-energy-storage-caes

[120] Using CO2 as energy storage https://cmte.ieee.org/futuredirections/2022/05/11/using-co2-as-energy-storage/

[121] Science Direct – Mechanical Energy Storage, Odne Stokke Burheim. Engineering Energy Storage 2017
https://www.sciencedirect.com/topics/engineering/mechanical-energy-storage

[122] IEEE – Environmental impacts of utility-scale battery storage in California. A. Balakrishnan *et al.*, "Environmental Impacts of Utility-Scale Battery Storage in California," *2019 IEEE 46th Photovoltaic Specialists Conference (PVSC)*, 2019, pp. 2472-2474, doi: 10.1109/PVSC40753.2019.8980665. https://ieeexplore.ieee.org/document/8980665/citations#citations

[123] RELiON – How Lithium Iron Phosphate batteries are easier on the environment
https://relionbattery.com/blog/how-lithium-iron-phosphate-batteries-are-easier-on-the-environment

[124] UL Research Institutes – Environmental impacts of lithium-ion batteries, March 16, 2022
https://ul.org/research/electrochemical-safety/getting-started-electrochemical-safety/environmental-impacts

[125] United States Environmental Protection Agency – Application of life-cycle assessment to nanoscale technology: Lithium-ion batteries for electric vehicles, April 24, 2013
https://www.epa.gov/sites/default/files/2014-01/documents/lithium_batteries_lca.pdf

[126] Croft Production Systems – Natural Gas Composition
https://www.croftsystems.net/oil-gas-blog/natural-gas-composition/

[127] Fact Sheet: Natural Gas Greenhouse Gas emissions
https://www.energy.gov/sites/prod/files/2014/07/f18/20140729%20DOE%20Fact%20sheet_Natural%20Gas%20GHG%20Emissions.pdf

[128] Department of Energy, United States of America – Liquified Natural Gas: Understanding the basic facts
https://www.energy.gov/sites/prod/files/2013/04/f0/LNG_primerupd.pdf

[129] Natural Gas, Addy Mettrick and Damon Ying, oil and gas statistics Dept of BEIS
https://assets.publishing.service.gov.uk/government/uploads/system/uploads/attachment_data/file/1006628/DUKES_2021_Chapter_4_Natural_gas.pdf

[130] Upstream carbon emissions: LNG vs pipeline gas – Wood Mackenzie April 2017

[131] Carbon footprint of global natural gas supplies to China. Yu Gan, Hassan M. El-Houjeiri, Alhassan Badahdah, Zifeng Lu, Hao Cai, Steven Przesmizki & Michael Wang

[132] Upstream carbon emissions: LNG vs pipeline gas – Wood Mackenzie April 2017

[133] Union of Concerned Scientists – Environmental impacts of natural gas
https://www.ucsusa.org/resources/environmental-impacts-natural-gas

[134] METGROUP – Natural gas environmental impact: Problems and Benefits, September 30, 2020
https://group.met.com/en/mind-the-fyouture/mindthefyouture/natural-gas-environmental-impact

[135] The carbon monoxide and gas safety society, Statistics of deaths and injuries 25/01/22
https://www.co-gassafety.co.uk/information/co-gas-safetys-statistics-of-deaths-and-injuries/

[136] Energy Saving Trust - Biomass https://energysavingtrust.org.uk/advice/biomass/

[137] Dept of BEIS - Digest of UK Energy Statistics, Annual data for UK, 2021
https://assets.publishing.service.gov.uk/government/uploads/system/uploads/attachment_data/file/1060151/DUKES_2021_Chapters_1_to_7.pdf

[138] Dept for BEIS – Biomass Policy Statement November 2021
https://assets.publishing.service.gov.uk/government/uploads/system/uploads/attachment_data/file/1031057/biomass-policy-statement.pdf

[139] Energy Saving Trust – Biomass. https://energysavingtrust.org.uk/advice/biomass/

Combustion-Free Heat, Power, and Energy

[140] Carbon debt and carbon sequestration parity in forest bioenergy production – Mitchell, Harmon, O'Connell

[141] Ecotricity Green Gas   https://www.ecotricity.co.uk/our-green-energy/green-gas

[142] CarbonBrief Clear on Climate Prof Sir John Beddington – Guest Post: Bioenergy 'flaw' under EU renewable target could raise emissions
https://www.carbonbrief.org/guest-post-bioenergy-flaw-under-eu-renewable-target-could-raise-emissions

[143] BBC News – The controversy of wood pellets as a green energy source 11 January 2021
https://www.bbc.co.uk/news/business-59546278

[144] BiofuelWatch – Coal-to-biomass conversions: Suplementing one (climate) disaster with another?
https://www.biofuelwatch.org.uk/wp-content/uploads/Coal-to-biomass-conversions.pdf

[145] CarbonBrief Clear on Climate Prof Sir John Beddington – Guest Post: Bioenergy 'flaw' under EU renewable target could raise emissions
https://www.carbonbrief.org/guest-post-bioenergy-flaw-under-eu-renewable-target-could-raise-emissions

[146] Centre for Sustainable Energy – Biomass heating   https://www.cse.org.uk/advice/renewable-energy/biomass-heating

[147] Which News Heating and Energy – Study says wood-burning stoves should be accompanied by a clear health warning 15 January 2021
https://www.which.co.uk/news/2021/01/study-says-wood-burning-stoves-should-be-accompanied-by-a-clear-health-warning/

[148] theBMJ – Air pollution in UK: the public health problem that won't go away *BMJ* 2015;350:h2757
https://www.bmj.com/content/350/bmj.h2757/rr-0

[149] CO2 purification and liquefaction plants
https://www.linde-engineering.com/en/process-plants/co2-plants/co2-purification-and-liquefaction/index.html

[150] Quanlin Zhou and Jens T. Birkholzer; On scale and magnitude of pressure build-up induced by large-scale geologic storage of CO_2; *Greenhouse Gases: Science & Technology*; 2010; DOI: 10.1002/ghg3 - https://www.soci.org/en/news/press/press_ghg_feb11

[151] Health and Safety Executive – Major hazard potential of CCS   https://www.hse.gov.uk/carboncapture/major-hazard.htm

[152] Plymouth Energy Community – Chelson Meadow Community Solar
https://plymouthenergycommunity.com/our-work/chelson-meadow

[153] Regen Plymouth waterfront decarbonisation feasibility study and energy plan
https://www.regen.co.uk/project/plymouth-waterfront-decarbonisation-feasibility-study-and-energy-plan/

[154] Office for National Statistics ONS Wind Energy in the UK: June 2021
https://www.ons.gov.uk/economy/environmentalaccounts/articles/windenergyintheuk/june2021

[155] The renewable energy hub – UK Onshore wind farms and wind speed interactive map
https://www.renewableenergyhub.co.uk/blog/uk-onshore-wind-farms-wind-speed-interactive-map/

[156] England's wind energy potential remains untapped – The Guardian 19 October 2021
https://www.theguardian.com/news/2021/oct/19/england-wind-energy-potential-onshore-windfarms-planning-development

[157] RenewableUK – The onshore wind industry prospectus October 2021
https://cdn.ymaws.com/www.renewableuk.com/resource/resmgr/media/onshore_wind_prospectus_fina.pdf

[158] Department for International Trade – Offshore wind
https://www.great.gov.uk/international/content/investment/sectors/offshore-wind/

[159] Department for International Trade – Offshore wind
https://www.great.gov.uk/international/content/investment/sectors/offshore-wind/

[160] RenewableUK – Press Release – 22 March 2022 – Offshore wind pipeline surges to 86 gigawatts, boosting UK's energy independence,
https://www.renewableuk.com/news/599739/Offshore-wind-pipeline-surges-to-86-gigawatts-boosting-UKs-energy-independence.htm

[161] Vindby – A Serious Offshore Wind Farm Design Game – Dornhelm, Seyr, Muskulus

[162] Scottish Government Statement 18 January 2022 – Scotwind offshore wind easing round: statement by the net zero Secretary
https://www.gov.scot/publications/scotwind-offshore-wind-leasing-round-statement-net-zero-secretary-18-january-2022/

[163] Offshore Wind Scotland   https://www.offshorewindscotland.org.uk/news-events/2022/january/scotwind-1-results/

[164] GE – Haliade-X offshore wind turbine – record setting offshore wind technology
https://www.ge.com/renewableenergy/wind-energy/offshore-wind/haliade-x-offshore-turbine

Climate Action Plymouth – Sustainable Power and Energy Group

[165] Office of energy efficiency and renewable energy – Wind Turbines: the bigger, the better.
https://www.energy.gov/eere/articles/wind-turbines-bigger-better

[166] All Sustainable Solutions – Vortex Bladeless Wind Turbine 22 July 2020 Rodoshi Das
http://www.wind-works.org/cms/index.php?id=399&tx_ttnews[tt_news]=2194&cHash=d1b21f3bd1f35d9e4804f1598b27bd86

[167] Aeromine Technologies https://www.aerominetechnologies.com

[168] All Sustainable Solutions – Vortex Bladeless Wind Turbine 22 July 2020 Rodoshi Das
https://allsustainablesolutions.com/vortex-bladeless-wind-turbine/

[169] The Guardian 1 June 2015 Can bladeless wind turbines mute opposition?
https://www.theguardian.com/sustainable-business/2015/jun/01/can-bladeless-wind-turbines-mute-opposition

[170] WEPD – The latest developments in wind turbine technology to facilitate radar and windfarm coexistence 16 February 2021
https://www.windpowerengineering.com/the-latest-developments-in-wind-turbine-technology-to-facilitate-radar-and-windfarm-coexistence/

[171] TurbineGenerator – Disadvantages of wind energy https://www.turbinegenerator.org/wind/advantages/disadvantages/

[172] The potential health impact of wind turbines – Chief Medical Officer of Health (CMOH) Report May 2010
https://health.gov.on.ca/en/common/ministry/publications/reports/wind_turbine/wind_turbine.pdf

[173] White & Case 26 April 2019 Insight https://www.whitecase.com/publications/insight/united-kingdom

[174] BMC Aquatic Biosystems 14 September 2014 Assessing environmental impacts of offshore wind farms: lessons learned and recommendations for the future https://aquaticbiosystems.biomedcentral.com/articles/10.1186/2046-9063-10-8

[175] EARTH.ORG Offshore wind farms: An ecological problem or environmental solution?
https://earth.org/offshore-wind-farms-an-ecological-problem-or-environmental-solution/

[176] Danish Energy Authority – Offshore wind farms and the environment – Danish experiences from Horns Rev and Nysted
https://naturstyrelsen.dk/media/nst/Attachments/havvindm_korr_16nov_UK.pdf

[177] BBC News 11 October 2021 Brown crabs find underwater power cables 'difficult to resist'.
https://www.bbc.co.uk/news/uk-scotland-south-scotland-58869465

[178] Northwest Renewable Energy Institute – Recycling wind turbine blades March 30, 2021 Mary Laurence
https://www.nw-rei.com/2021/03/30/recycling-turbine-blades/

[179] The Guardian 23 August 2022 Wind turbine blades could be recycled into gummy bears, scientists say – Chelsie Henshaw
https://www.theguardian.com/environment/2022/aug/23/wind-turbine-blades-could-recycled-gummy-bears-scientists

[180] List.Solar Largest solar power stations in UK 01 June 2021 https://list.solar/plants/largest-plants/uk/

[181] S&P Global Commodity Insights 29 May 2020 UK gives consent for 350MW Cleve Hill solar farm https://www.spglobal.com/commodity-insights/en/market-insights/latest-news/electric-power/052920-uk-gives-consent-for-350-mw-cleve-hill-solar-farm

[182] Cleve Hill Solar Park https://www.clevehillsolar.com

[183] Clean Energy Reviews Most powerful solar panels 2022 July 19, 2022 - Jason Svarc
https://www.cleanenergyreviews.info/blog/most-powerful-solar-panels

[184] BSG:Ecology 10 April 2019 Impacts of operational solar farms on biodiversity: a review of studies and call for evidence
https://www.bsg-ecology.com/impacts-of-operational-solar-farms-on-biodiversity/

[185] BSG:Ecology Potential ecological impacts of ground mounted photovoltaic solar panels – An introduction and literature review April 2019 https://www.bsg-ecology.com/wp-content/uploads/2019/04/Solar-Panels-and-Wildlife-Review-2019.pdf

[186] Green Building Press – Can solar farms be good for biodiversity 4 September 2013
https://www.greenbuildingpress.co.uk/article.php?category_id=1&article_id=1549

[187] BayWa r.e. Agri-PV – Together for more Climate Resilience https://www.baywa-re.com/en/solar-projects/agri-pv

[188] Intelligent Living Grazing sheep under solar panels rids of weeds and offers wool – Luana Steffen 10 September 2020
https://www.intelligentliving.co/grazing-sheep-under-solar-panels/

[189] E&T April 30 2021 Mixing sheep grazing with solar farms yields major land productivity increase.

Combustion-Free Heat, Power, and Energy

https://eandt.theiet.org/content/articles/2021/04/mixing-sheep-grazing-with-solar-farms-yields-major-land-productivity-increase/

[190] Planning guidance for the development of large scale ground mounted solar PV systems – bre national solar centre
https://www.bre.co.uk/filelibrary/pdf/other_pdfs/KN5524_Planning_Guidance_reduced.pdf

[191] The energyst – DEFRA "considering solar ban on lower quality English farmland". https://theenergyst.com/defra-considering-ban-on-solar-farms-on-lower-quality-english-farmland/

[192] PagerPower Urban & Renewables – The UK's floating photovoltaic (FPV) potential October 28, 2020 – Michael Sutton
https://www.pagerpower.com/news/the-uks-floating-photovoltaic-fpv-potential,'

[193] Plymouth City Council – Planning submitted for new solar farm 21 February 2022
https://www.plymouth.gov.uk/newsroom/pressreleases/planningsubmittednewsolarfarm

[194] Oceans of Energy – A world's first: offshore floating solar farm installed at the Dutch North Sea. 11 December 2019 https://oceansofenergy.blue/2019/12/11/a-worlds-first-offshore-floating-solar-farm-installed-at-the-dutch-north-sea/

[195] Marine Technology News July 1, 2022 – Portugal's EDP to scale up offshore solar plants in Southeast Asia
https://www.marinetechnologynews.com/news/portugal-scale-offshore-solar-620857

[196] National Library of Medicine – National Center for Biotechnology information November 16 - Environmental impacts of solar photovoltaic systems: A critical review of recent progress and future outlook. https://pubmed.ncbi.nlm.nih.gov/33234276/

[197] AZO Cleantech – The environmental impacts of photovoltaic technology – Dr Liji Thomas, MD Jan 3, 2019
https://www.azocleantech.com/article.aspx?ArticleID=831

[198] Clarkson & Woods Ecological Consultants – The effects of solar farms on local biodiversity: A comparative study 2015
https://www.clarksonwoods.co.uk/news/news_solarresearch.html

[199] H. Montag, G Parker & T Clarkson 2016 The Effects of Solar Farms on Loacl Biodiversity; A Comparitive Study. Clarkson and Woods and Wychwood Biodiversity.

[200] Bat Conservation Trust – Our vision is of a world rich in wildlife where bats and people thrive together https://www.Bats.org.uk

[201] BSG:ecology – Potential ecological impacts of ground-mounted photovoltaic solar panels – An introduction and literature review. April 2019 https://www.bsg-ecology.com/wp-content/uploads/2019/04/Solar-Panels-and-Wildlife-Review-2019.pdf

[202] Government Guidance – Harnessing hydroelectric power. How hydroelectric power works, regional schemes, and information on installing your own micro-hydro scheme. 22 January 2013 https://www.gov.uk/guidance/harnessing-hydroelectric-power

[203] Office of Energy Efficiency and Renewable Energy – Water-Power Technologies Office – Benefits of Hydropower
https://www.energy.gov/eere/water/benefits-hydropower

[204] BusinessLive – New plans for a Severn Barrage generating 10% of the UK's electricity needs – Sion Barry, 11 March 2022.
https://www.business-live.co.uk/economic-development/new-plans-severn-barrage-generating-23356609

[205] TheGreenAge – Should we build the Severn Barrage? – Surely a no brainer – June 10, 2013
https://www.thegreenage.co.uk/should-we-build-the-severn-barrage-surely-a-no-brainer/

[206] energysage – Environmental impacts of hydropower 27 09 2019
https://www.energysage.com/about-clean-energy/hydropower/environmental-impacts-hydropower/

[207] What are the environmental impacts of hydropower – Jennifer Okafor 08:03:22
https://www.trvst.world/renewable-energy/environmental-impacts-of-hydropower/

[208] TurbineGenerator – Environmental Impact of Hydroelectricity
https://www.turbinegenerator.org/hydro/environmental-impact-hydroelectricity/'

[209] Government press release – UK government announces biggest investment into Britain's tidal power 24 November 2021
https://www.gov.uk/government/news/uk-government-announces-biggest-investment-into-britains-tidal-power

[210] NS Energy MeyGen Tidal Power Project, Pentland Firth
https://www.nsenergybusiness.com/projects/meygen-tidal-power-project/#

[211] Marine Energy Wales – Education Programme https://www.marineenergywales.co.uk/about/education/

[212] Marine Scotland Assessment – Case study: Nova Innovation – Shetland Tidal Array http://marine.gov.scot/sma/assessment/case-study-nova-innovation-shetland-tidal-array

Climate Action Plymouth – Sustainable Power and Energy Group

[213] Orbital Marine Power – Orbital Marine Power awarded two CfDs as part of UK Government renewable energy auction
https://orbitalmarine.com/orbital-awarded-two-cfds/

[214] EnFait Enabling Future Arrays in Tidal https://www.enfait.eu

[215] Tocardo – What is the impact of tidal energy on the environment? 3 July 2020 Andries van Unen
https://www.tocardo.com/what-is-the-impact-from-tidal-power-on-the-environment/

[216] Examining the effectiveness of support for UK wave energy innovation since 2000: Lost at sea or a new wave of innovation? Hannon, Matthew and van Diemen, Renée and Skea, Jim (2017) (https://doi.org/10.17868/62210) https://strathprints.strath.ac.uk/62210/

[217] BBC News – Wave power firm Pelamis calls in adfministrators 21 November 2014
https://www.bbc.co.uk/news/uk-scotland-scotland-business-30151276

[218] The Engineer – Leading wave energy pioneer Prof Stephen Salter
https://www.theengineer.co.uk/leading-wave-energy-pioneer-prof-stephen-salter/

[219] electricity info – Wave Power 24 September 2022 https://electricityinfo.org/news/wave-power-49/

[220] Europewave – Europwave's successful wave energy projects unveiled 7 December 2021
https://www.europewave.eu/news/europewaves-successful-wave-energy-projects-unveiled

[221] https://www.researchgate.net/figure/Pelamis-Figure-2-Wave-Dragon_fig1_329628734

[222] The Engineer – Mocean Energy unveils Blue X wave energy converter
https://www.theengineer.co.uk/mocean-energy-blue-x-wave-energy-converter/

[223] BBC News – Cornwall wave hub to be sold for offshore wind farm
https://www.bbc.co.uk/news/uk-england-cornwall-57156482

[224] Eco Wave Power https://www.ecowavepower.com/about/who-are-we/

[225] Eco Wave Power – Eco Wave Power presents continued operational progress and reports first half 2022 financial results 27 September 2022 https://www.ecowavepower.com/eco-wave-power-presents-continued-operational-progress-and-reports-first-half-2022-financial-results/

[226] Alternative Energy Tutorial – Wave energy uses the power of the waves
https://www.alternative-energy-tutorials.com/wave-energy/wave-energy.html

[227] British Geological Survey – Geothermal energy https://www.bgs.ac.uk/geology-projects/geothermal-energy/

[228] TownRock Energy – Geothermal energy in the UK – What, why, where and how.
http://townrockenergy.com/wp-content/uploads/2020/07/TownRock-brochure-2020.pdf

[229] https://geothermal-energy-journal.springeropen.com/articles/10.1186/s40517-016-0046-8#Fig1

[230] Geothermal Energy – 16 June 2017. Assessment of the resource base for engineered geothermal systems in Great Britain
https://geothermal-energy-journal.springeropen.com/articles/10.1186/s40517-017-0066-z

[231] JPT – Geothermal – DOE offers grants to advance geothermal drilling speed February 14, 2022 Blake Wright
https://jpt.spe.org/doe-offers-grants-to-advance-geothermal-drilling-speed

[232] Southampton's geothermal pursuit: A success or failure
https://www.wessexscene.co.uk/magazine/2017/11/30/southamptons-geothermal-pursuit-a-success-or-failure/

[233] Eden Geothermal – Eden geothermal hot rocks project hits target with successful drilling of first well
https://www.edengeothermal.com/press/eden-geothermal-hot-rocks-project-hits-target-with-successful-drilling-of-first-well/

[234] BBC News – Seismic activity stops geothermal drilling at Eden Project https://www.bbc.co.uk/news/uk-england-cornwall-60689204

[235] ITVNews Earthquake sparked by Eden Project's geothermal testing 'shakes homes' in Cornwall.
https://www.itv.com/news/westcountry/2022-03-10/it-shook-my-house-earthquake-felt-from-within-homes-in-cornwall

[236] Eden Geothermal Well Testing https://www.edengeothermal.com/the-project/drilling-and-operations/well-testing/

[237] United Downs – An overview of the United Downs Deep Geothermal Power (UDDGP) Project https://www.marriottdrilling.com/wp-content/uploads/2019/08/An-Overview-of-the-United-Downs-Deep-Geothermal-Power-UDDGP-Project.pdf

[238] Geothermal engineering develops projects for 20MW of power in Cornwall July 2, 2021

Combustion-Free Heat, Power, and Energy

https://renewablesnow.com/news/geothermal-engineering-develops-projects-for-20-mw-of-power-in-cornwall-746392/

[239] The Coal Authority – Geothermal energy from abandoned coal mines
https://www2.groundstability.com/geothermal-energy-from-abandoned-coal-mines/

[240] Groundsure – Urban mining in and around Plymouth https://www.groundsure.com/urban-mining-in-and-around-plymouth/

[241] Union of Concerned Scientists – Environmental impacts of geothermal energy – Mar 5, 2013
https://www.ucsusa.org/resources/environmental-impacts-geothermal-energy

[242] Alternative Energy Tutorials – Geothermal Energy
https://www.alternative-energy-tutorials.com/geothermal-energy/geothermal-energy.html

[243] Mladen BOŠNJAKOVIĆ et al.: Environmental Impact of Geothermal Power Plants

[244] Hansard Plutonium Production https://api.parliament.uk/historic-hansard/written-answers/1981/jul/24/plutonium-production

[245] Government News Story – Decommissioning agreement reached on advanced gas cooled reactor (AGR) nuclear power stations 23 June 2021 https://www.gov.uk/government/news/decommissioning-agreement-reached-on-advanced-gas-cool-reactor-agr-nuclear-power-stations

[246] https://inis.iaea.org/collection/NCLCollectionStore/_Public/29/010/29010110.pdf

[247] New Civil Engineer MPs call for decommissiong delay for UK's ageing nuclear power stations 20 May 2022 Rob Hakimian
https://www.newcivilengineer.com/latest/mps-call-for-decommissioning-delay-for-uks-ageing-nuclear-power-stations-20-05-2022/

[248] BBC News - Hinkley Point C nuclear plant to open later at greater cost 27 January 2021
https://www.bbc.co.uk/news/uk-england-somerset-55823575

[249] BBC News Hinkley Point C delayed by a year as cost goes up by £3bn
https://www.bbc.co.uk/news/uk-england-somerset-61519609.amp

[250] World Nuclear News 12 January 2022 – Fresh delay to Flamanville 3 blamed on pandemic.
https://www.world-nuclear-news.org/Articles/Fresh-delay-to-Flamanville-blamed-on-impact-of-pan

[251] Government Collection Hinkley Point C – Contracts for the Hinkley Point C which from 2025 is expected to provide 3.2GW of clean, reliable, electricity generation capacity. 17th July 2018   https://www.gov.uk/government/collections/hinkley-point-c

[252] Catalyst Digital Energy - Wholesale Electricity Prices. https://www.catalyst-commercial.co.uk/wholesale-electricity-prices/

[253] Government Collection Hinkley Point C – Contracts for the Hinkley Point C which from 2025 is expected to provide 3.2GW of clean, reliable, electricity generation capacity. 17th July 2018   https://www.gov.uk/government/collections/hinkley-point-c

[254] National Infrastructure Planning – The Sizewell C Project
https://infrastructure.planninginspectorate.gov.uk/projects/eastern/the-sizewell-c-project/

[255] BBC News Sizewell C nuclear plant campaigners challenge approval 03 August.
https://www.bbc.co.uk/news/uk-england-suffolk-62467027?at_campaign=64&at_custom1=link&at_custom4=086122F2-1730-11ED-9BDD-F58B4744363C&at_medium=custom7&at_custom2=twitter

[256] BEIS electricity statistics
https://assets.publishing.service.gov.uk/government/uploads/system/uploads/attachment_data/file/1006701/DUKES_2021_Chapter_5_Electricity.pdf

[257] Government UK nuclear: Powering the future 26 January 2015
https://www.gov.uk/government/publications/uk-nuclear-powering-the-future/uk-nuclear-powering-the-future

[258] World Nuclear Association – World nuclear power reactors & uranium requirements Sept 2022 https://www.world-nuclear.org/information-library/facts-and-figures/world-nuclear-power-reactors-and-uranium-requireme.aspx

[259] France announces plans to build up to 14 nuclear reactors 11 February 2022
https://edition.cnn.com/2022/02/11/business/nuclear-power-france/index.html

[260] Republic World – South Korea halts programme to gradually phase out nuclear energy https://www.republicworld.com/world-news/rest-of-the-world-news/south-korea-halts-programme-to-gradually-phase-out-nuclear-energy-report-articleshow.html

[261] BBC News Energy strategy: UK plans eight new nuclear reactors to boost production 7 April 2022
https://www.bbc.co.uk/news/business-61010605

Climate Action Plymouth – Sustainable Power and Energy Group

[262] New Atlas – US nuclear regulator greenlights its first small modular reactor. https://newatlas.com/energy/nrc-certifies-nuscale-nuclear/?fbclid=IwAR36Z3bPQYXwKx10J4l9ZXwh1dJk13QXCjp85_UtQI4vRB7i9cTsqfTChM4

[263] Rolls-Royce Nuclear – Small Modular Reactors – once in a lifetime opportunity for the UK https://www.rolls-royce.com/~/media/Files/R/Rolls-Royce/documents/customers/nuclear/smr-brochure-july-2017.pdf

[264] The Engineer – Rapid reaction: small factory-built nuclear reactors could be delivered by lorry 13 March 2017 https://www.theengineer.co.uk/rapid-reaction-the-small-factory-built-nuclear-reactors-could-be-delivered-on-the-back-of-a-lorry/

[265] The Guardian – Rolls-Royce expects UK approval for small nuclear reactors by mid-2024 – Jasper Jolly 19 April 2022 https://www.theguardian.com/business/2022/apr/19/rolls-royce-expecting-uk-approval-for-small-nuclear-reactors-by-mid-2024

[266] Capital Com Rolls-Royce: New SMR tech "a decarbonisation gamechanger" https://capital.com/rolls-royce-new-smr-tech-a-decarbonisation-gamechanger

[267] Proceedings of the National Academy of Sciences PNAS - Nuclear waste from small modular reactors, Lindsay M. Krall, Allison M. MacFarlane, and Rodney C. Ewing May 31, 2022 https://news.stanford.edu/2022/05/30/small-modular-reactors-produce-high-levels-nuclear-waste/

[268] Rolls-Royce announces funding secured for Small Modular Reactors https://www.rolls-royce.com/media/press-releases/2021/08-11-2021-rr-announces-funding-secured-for-small-modular-reactors.aspx

[269] The Engineer Expert Q&A: small modular reactors https://www.theengineer.co.uk/expert-qa-small-modular-reactors/

[270] EGB Engineering – The UK Government's Development of Advanced Modular Reactors (AMR's) https://egb-eng.com/the-uk-governments-development-of-advanced-modular-reactors-amrs/

[271] Moltex Clean Energy Moltex successful in the UK government £300k Advanced Modular Reactor competition https://www.moltexenergy.com/moltex-successful-in-the-uk-government-300k-advanced-modular-reactors-competition/

[272] BBC 39 ways to save the planet https://www.bbc.co.uk/programmes/m000z79l

[273] Moltex Clean Energy Moltex successful in the UK government £300k Advanced Modular Reactor competition https://www.moltexenergy.com/moltex-successful-in-the-uk-government-300k-advanced-modular-reactors-competition/

[274] NS Energy Stable Salt Reactor: New concept for nuclear waste-burning Ian Scott and Simon Newton 11 April 2022 https://www.nsenergybusiness.com/features/stable-salt-reactor-new-concept-for-nuclear-waste-burning/

[275] NS Energy Stable Salt Reactor: New concept for nuclear waste-burning Ian Scott and Simon Newton 11 April 2022 https://www.nsenergybusiness.com/features/stable-salt-reactor-new-concept-for-nuclear-waste-burning/

[276] Moltex Clean Energy – Moltex encouraged by SMR feasibility report https://www.moltexenergy.com/moltex-encouraged-by-smr-feasibility-report/

[277] Moltex Clean Energy – Moltex encouraged by SMR feasibility report https://www.moltexenergy.com/moltex-encouraged-by-smr-feasibility-report/

[278] Next Big Future - Moltex molten salt reactor being built in New Brunswick Canada Brian Wang July 19, 2018 https://www.nextbigfuture.com/2018/07/moltex-molten-salt-reactor-being-built-in-new-brunswick-canada.html

[279] National Library of Medicine – Potential Human Health Effects of Uranium Mining, Processing, and Reclamation. https://www.ncbi.nlm.nih.gov/books/NBK201047/

[280] World Nuclear Association – Safety of nuclear power reactors March 2022 https://www.world-nuclear.org/information-library/safety-and-security/safety-of-plants/safety-of-nuclear-power-reactors.aspx

[281] Friends of the Earth Policy – Nuclear energy: our position. 27 November 2017 https://policy.friendsoftheearth.uk/policy-positions/nuclear-energy-our-position

[282] UK Parliament – House of Lords library – In Focus, Nuclear power in the UK - Emily Hoves 01 December 2021 https://lordslibrary.parliament.uk/nuclear-power-in-the-uk/

[283] Cost estimate of Olkiluoto disposal facility for spent nuclear fuel
Kukkola, T. (Fortum Nuclear Services Ltd, Espoo (Finland)); Saanio, T. (Saanio and Riekkola Oy, Helsinki (Finland))
Posiva Oy, Helsinki (Finland)
https://inis.iaea.org/search/search.aspx?orig_q=RN:36090684

[284] Reuters – 18 August 2021 Francois Murphy – Iran accelerates enrichment of uranium to near weapons-grade, IAEA says https://www.reuters.com/world/middle-east/iran-accelerates-enrichment-uranium-near-weapons-grade-iaea-says-2021-08-17/

Combustion-Free Heat, Power, and Energy

[285] BBC News 9 February 2022 – Major breakthrough on nuclear fusion energy
https://www.bbc.co.uk/news/science-environment-60312633

[286] Plymouth City Council – Briefing Report – City Centre district energy scheme
https://democracy.plymouth.gov.uk/documents/s93087/Part%20ONE%20Briefing%20Report.pdf

[287] Regen transforming energy – Plymouth waterfront decarbonisation feasibility study and energy plan
https://www.regen.co.uk/project/plymouth-waterfront-decarbonisation-feasibility-study-and-energy-plan/

[288] Energy Saving Trust – 13 September 2021 What is district heating? A low carbon solution for the UK's homes.
https://energysavingtrust.org.uk/what-district-heating/

[289] IEA District Heating September 2022 https://www.iea.org/reports/district-heating

[290] Polar Night Energy = Store wind and solar power as heat in sand https://polarnightenergy.fi/technology

[291] IEA District Heating September 2022   https://www.iea.org/reports/district-heating

[292] Government Collection – Heat networks investment project (HNIP): overview and how to apply
https://www.gov.uk/government/collections/heat-networks-investment-project-hnip-overview-and-how-to-apply

[293] Government Collection – Heat Networks Investment Project (HNIP): overview and how to apply.
https://www.gov.uk/government/collections/heat-networks-investment-project-hnip-overview-and-how-to-apply

[294] Triple Point Heat Networks Investment Management 0 Some of our successful HNIP funded projects
https://tp-heatnetworks.org/funded-projects/

[295] The Association for Decentralised Energy – Bringing Energy Together – Market Report: Heat Networks in the UK
https://www.theade.co.uk/assets/docs/resources/Heat%20Networks%20in%20the%20UK_v5%20web%20single%20pages.pdf

[296] 220504 Paris- Plymouth pilot update.pdf

[297] Plymouth's carbon saving energy upgrade public buildings https://www.plymouth.gov.uk/carbon-saving-energy-upgrade-public-buildings.

[298] Environmental impacts of the infrastructure for district heating in urban neighbourhood, Jordi Oliver-Solà XavierGabarrell[ac]JoanRieradevall[ac] November 2009 https://www.sciencedirect.com/science/article/abs/pii/S0301421509004418

[299] Plymouth Live - Massive green hydrogen hub planned for Langage Energy Park 1 September 2022
https://www.plymouthherald.co.uk/news/plymouth-news/massive-green-hydrogen-hub-planned-7534496?

[300] Carlton Power – Carlton Power plans to build Devon and Cornwall's first low carbon hydrogen hub 1 September 2022
https://www.carltonpower.co.uk/news/carlton-power-plans-to-build-devon-amp-cornwalls-first-low-carbon-hydrogen-hub

[301] Rosenow, Is heating homes with hydrogen all but a pipe dream? An evidence review, Joule (2022),
https://doi.org/10.1016/j.joule.2022.08.015

[302] BBC News – Study contradicts Rees-Mogg over hydrogen for heating
https://www.bbc.co.uk/news/science-environment-63050910?at_medium=RSS&at_campaign=KARANGA

[303] Cornwall Insight – Hydrogen Costs 22 September 2022 https://www.cornwall-insight.com/wp-content/uploads/2022/09/MCS-Insight-Paper-Hydrogen-Sept-2022.pdf?utm_source=website&utm_medium=website

[304] Office of Energy Efficiency & Renewable Energy – Fuel Cell Basics https://www.energy.gov/eere/fuelcells/fuel-cell-basics

[305] The Guardian Pollutionwatch: Olympic flame is a warning sign for hydrogen future – Gary Fuller 13 August 2021
(https://www.theguardian.com/environment/2021/aug/13/pollutionwatch-olympic-flame-warning-sign-hydrogen-future

Printed in Great Britain
by Amazon